LICHEN T

FROM THE ALLEGHANIES

LICHEN TUFTS,

FROM THE ALLEGHANIES

Elizabeth C. Wright

With a New Introduction by Emily E. VanDette

Afterword by Laurie Lounsberry Meehan

EXCELSIOR EDITIONS

Published by State University of New York Press, Albany
© 2022 State University of New York

Excelsior Editions is an imprint of State University of New York Press

For information, contact State University of New York Press, Albany, NY
www.sunypress.edu
Library of Congress Cataloging-in-Publication Data

Names: Wright, Elizabeth C., author
With a new introduction by Emily E. VanDette
Title: Lichen tufts, from the Alleghanies
Description: Albany : State University of New York Press, [2022] | Series:
    New York Classics | Excelsior Editions
Identifiers: ISBN 9781438489216 (hardcover : alk. paper) | ISBN
    9781438489223 (e-book) | ISBN 9781438489209 (paperback)
Further information is available at the Library of Congress.

10 9 8 7 6 5 4 3 2 1

Somebody says that "A Cathedral would hardly hold my acquaintances-the pulpit would accommodate my friends." 'Ellis volume, with my compliments, is addressed to the Cathedral full-to the few in the pulpit it is dedicated with the love of

A. C. WHEELER

CONTENTS.

NEW INTRODUCTION.

Emily E. VanDette

Part environmental manifesto, part poetry collection, part *Walden*-inspired ode to the Allegheny wilderness, the 1860 book *Lichen Tufts, from the Alleghanies* defies easy classification. The same can be said for its author. In some ways, Elizabeth C. Wright was a woman ahead of her time: a professional scientist, teacher, lecturer, and activist for gender and racial equality and social justice; an independent and intellectual woman who did not follow the expected path for a woman to marry young, raise a family, and devote herself to a domestic life. In other ways, Wright's life and writing reflect the circumstances and environments of her upbringing, education, and political and social networks in western New York and beyond. She was indeed the perfect candidate to author the first known book-length treatise on nature written by a US woman. Wright's world was one of incredible social turmoil and change, spiritual and cultural awakenings, and a new sense of urgency around the intertwining issues of environmental and social consciousness. The beauty and wisdom contained in *Lichen Tufts*, and the remarkable life story of its author, are more relevant today than ever.

Lichen Tufts consists of four essays about nature followed by forty eclectic poems. The first essay, "Into the Woods," documents a camping trip in the vicinity of the present-day Allegany State Park and Seneca Nation territory, which the author took with a group of friends and mentors from her recent alma mater, Alfred University. The group of like-minded adventurers and nature lovers included a radically progressive young faculty couple, Jonathan and Abigail Allen, and three recent students and graduates (the cohort is discussed in detail in the afterword, written by Alfred University archivist Laurie Lounsberry Meehan). Taken on its own, "Into the Woods" narrates an exciting and remarkably modern-sounding adventure shared by a group of energetic, idealistic, and progressive young men and women. As the opening essay of

Lichen Tufts, it also serves as a practical, real-life basis for the philosophical discussions throughout the rest of the book. Not surprisingly, the idea of women camping out in the wilderness was met with resistance, but throughout the essays of *Lichen Tufts*, Wright develops a feminist response to the restrictions placed on women's physical and intellectual lives. Wright rejoices in the collective feeling of liberation experienced by her camping party: "It was utterly delightful to let ourselves loose, and live freely; to have no rules for coming in or going out, for rising up or sitting down; to be emancipated from the bondage of the ceremonial law, and do what pleased us best, was paradisiacal enough."[1] Drawing upon her education in natural science and philosophy, a rare background for a woman of her day, Wright expresses a transcendentalist appreciation for the natural world. "Into the Woods" highlights the ways in which her camping party broke free from gender roles and other social conventions, with the men in her party cooking meals and the women trading their conventional attire for practical hiking clothes and learning how to shoot a gun.

While clearly aligned with the transcendentalist movement, Wright's passion for the natural world and social justice can be traced back to her roots in a tightly knit, multigenerational Quaker family that lived on the border of New York and Pennsylvania.[2] Wright's grandparents, Robert and Elizabeth Clendenon, were missionaries at the Quaker-run Tunesassa school for the Seneca Nation in Cattaraugus County, New York. The Clendenons' daughters and their spouses stayed close to home, raising their families in a closely knit extended family. Abigail Clendenon and Asahel Wright, Elizabeth Wright's parents, were married in the Ceres, Pennsylvania, home of Abigail's sister Hannah and her husband John King on August 10, 1825. While the family lived too far from the nearest Quaker meetings to participate actively in their faith (and Abigail married a non-Quaker), they nevertheless continued to abide by Quaker values. Elizabeth Clendenon Wright was born in December 1826 in Ceres, and she and her sisters, Lydia Ellen and Sarah Ann, grew up in close proximity to their maternal grandparents, aunts, uncles, and cousins. The multigenerational family nurtured women's education and voice, religious faith, an appreciation for the natural world, and a passionate dedication to social justice and peace. For all of their otherwise progressive values and practices, the Clendenons' work as missionaries with the people of the Seneca Nation reflects

1   Elizabeth C. Wright, *Lichen Tufts, from the Alleghanies* (New York: M. Doolady, 1860), 17.
2   The biographical research about Wright is from Emily E. VanDette, "Elizabeth C. Wright," *Legacy: A Journal of American Woman Writers* 37. no.2 (2021).

an assimilation agenda connected to historical patterns of dominance and era-
sure of Native American communities and culture. While boarding schools like
Tunesassa may have been intended initially as a benevolent means of helping
Native American communities adapt to Euro-American culture, they became
a main tool in the US government's forced assimilation program in the late
nineteenth and early twentieth centuries. The autobiographical work "The
School Days of an Indian Girl," written in 1900 by Zitkala-Sa (Lakota Sioux),
provides a compelling firsthand account of the author's traumatic boarding
school experience. The legacy of her grandparents' missionary mindset mani-
fests in Wright's writing about her encounters with Seneca people in "Into the
Woods," especially in her approving commentary about the "advantages of true
civilization" for the Senecas. Even while advocating for and aiding oppressed
people, activists and missionaries often contributed to the normalization of
ideas responsible for generations of trauma and cultural loss.

The destructive effects of their assimilationist missionary work notwith-
standing, the Quakers were among the most radical activists in the campaign
to end slavery. Wright's immediate and extended family played an active role
in the antislavery cause, and they lived their day-to-day lives in accordance
with their beliefs. They participated in the free-produce movement, tapping
maple trees and boycotting sugar and other goods produced by slave labor, and
they provided shelter and aid to fugitives from slavery. The home of Elizabeth
Wright's maternal aunt and uncle Hannah and John King, where Wright and
her sisters surely spent a lot of time during their childhood, was a key "sta-
tion" in the Underground Railroad. The children of the Wrights and Kings
were clearly impacted by their upbringing in an environment of radical, fearless
antislavery activism. The daughter of the Kings, Mary, married a like-minded
abolitionist, John S. Mann, and the couple established Underground Railroad
stations in their home and in Mary's bookstore in Coudersport, Pennsylvania.
Elizabeth and her sister Lydia Ellen (often referred to as "Ellen" by family and
friends) were close to their cousin Mary King Mann throughout their entire
lives, and Ellen even lived with the Manns in their Coudersport home for several years. Carrying on with the activism modeled by their parents, the women
worked together on multiple areas of social reform and public service. They
contributed tirelessly to the temperance campaign, founded a public library,
worked for educational reform, and, of course, contributed to efforts to abolish
slavery. In addition to reinforcing Wright's interests in education and social
justice, this collaborative family network also encouraged women's participation in public issues.

The importance of education was a constant theme throughout Wright's life. As a teacher in the first public school in Keating, Pennsylvania, founded in 1858, she served on the boards of local teachers associations and helped shape policy and curriculum and broaden access to public education. She was joined in this work by her sister Ellen and their Coudersport cousins, the Manns. In addition to the influence of her family network, Wright found ideal role models and a progressive example of equitable education at Alfred. The young faculty couple Jonathan and Abigail Allen, who would become Wright's life-long friends, were dedicated to creating an inclusive coeducational environment, where women were trained in areas typically reserved for men, such as the sciences and public speaking. The Allens were also feminists who fought for women's equal pay and voting rights, and they were radical abolitionists and environmentalists. In addition to finding kindred spirits in the Allens, Wright felt perfectly at home at Alfred, where the overall atmosphere and curriculum were ahead of its time, and it surely influenced her own philosophy and commitment as an educator throughout the rest of her life. The afterword describes the climate and personalities of Wright's Alfred network in detail.

Growing up on the New York-Pennsylvania border in Ceres and educated at Alfred, Wright was undoubtedly influenced by the spiritual fervor and social radicalism that characterized the "Burned Over District" of western New York. Along with religious revivalism and the formation of Spiritualism and other experiments, the region between Albany and Buffalo was a hotbed of radical reform campaigns for temperance, abolitionism, and women's rights. Wright, like many progressive-minded, educated women of her day, played an active role in all of those movements. Unlike most of her female peers at the time, though, Wright was trained in public speaking at Alfred, which equipped her to become a prominent speaker in the temperance lecture circuit. At a time when women were still barred from speaking in venues throughout much of the country, Wright's presence on the stage was a novelty, and she took advantage of the opportunity to spread the message against the sale and consumption of alcohol to her fascinated audiences. In 1857, she became one of the first women to be nominated to an executive position in the American Temperance Union, a significant achievement given the usual barriers against women's full participation in such organizations. Because of the correlation between alcoholism and domestic violence, temperance was considered a "women's issue." Women who fought to ban or restrict the sale and consumption of alcohol often saw that campaign as their only course of action to prevent domestic violence, during a time when women were denied the right to vote and other basic rights of citizenship. When a woman married, she gave up much of her legal autonomy and control over her property and finances, and women were often denied custody of their children in the case of divorce, even if her husband was violent. The subordinate and dependent legal status of married women made them especially vulnerable to the effects of a husband's alcoholism, so an increasing number of women were motivated to participate in the grassroots temperance movement.

Women's participation—and perhaps, ironically, their exclusion—from the temperance campaign contributed to the momentum of the women's rights and suffrage movements at the time. Rural communities throughout western and central New York saw a surge of activism and support for the temperance cause, and while women worked at the ground level for the "crusade," as it was termed, they were typically excluded from the organizational infrastructure of the movement. That exclusion, combined with the growing realization that the only real recourse for women to protect themselves and their children was by acquiring legal autonomy and the right to vote, reinforced the crossover between the temperance and women's rights campaigns. A turning point in this history was the Whole World's Temperance Convention, held at the Metropolitan Hall in New York City in September 1853. In a dramatic display of opposition to women's participation, the men leading the convention refused to include the female delegates in attendance (including Susan B. Anthony and Lucy Stone). While a small number of men vocally supported the women and protested their expulsion from the convention, the majority objected to the women's participation and drowned out the voices of the women who attempted to speak. In making their case for removing the women from the convention, the men quoted scripture, charged the women delegates with "outraging the proprieties of [their] sex," and "referred to 'women in breeches' as a disgrace to their sex."[3] The exclusion of women from the convention served as a catalyst for the women's rights movement. The following month (October 1853), at a women's rights convention held in Cleveland, Ohio, the women discussed at length the unjust treatment several of their fellow activists had encountered at the event they dubbed the "Half World's Convention" in New York (because only half of the population was represented). Elizabeth C. Wright attended the women's rights convention in Cleveland and spoke in favor of a diplomatic approach to collaborating with like-minded temperance men. Like many progressive women of the era, she worked for multiple social reform causes while navigating the misogynistic biases against women's participation in public issues. Wright's

3 *Whole World's Temperance Convention, Lucy Stone, and National American Woman Suffrage Association Collection* (New York: Fowlers and Wells, 1853), 8.

nomination to an executive position for the American Temperance Union in 1857 represented an important milestone for women and a step toward women being accepted and acknowledged for their social reform service.

Contributing to the temperance cause empowered many women to extend their public participation in social reform to the most urgent issue of the day, the abolition of slavery. For Wright, her childhood upbringing, exposure to the radical atmosphere of the Burned Over District, and her unusually progressive education at Alfred combined for a remarkable set of influences on her abolitionist stance. Even among other antislavery activists, Wright was as outspoken and radical as they came. In December 1853, Wright attended a meeting of the American Anti-Slavery Society, along with such prominent abolitionist leaders as William Lloyd Garrison, Sojourner Truth, and Lucretia Mott. Wright's voice appears in multiple places in the proceedings of the meeting, including her poem "A Word to the Weary," which she later included in *Lichen Tufts*. Written as an occasional poem for the twenty-year anniversary of the American Anti-Slavery Society's founding, the poem offers a message of encouragement and hope to the activists who had fought for decades to end slavery. In addition to reading her poem, Wright gave impromptu remarks and engaged in a debate with a man in attendance at the meeting about the role of individual moral responsibility in the perpetuation of slavery. She argued, among other points, that everyone was morally accountable for slavery, and "there was a great and terrible responsibility resting upon all who allowed those heavy burden to be borne by others which they would not allow themselves to touch with their little fingers."[4] Wright's belief in the moral responsibility of abolitionism, and her self-possession in debating statements made by a man in a public meeting, both stem from the progressive atmosphere of her abolitionist family that encouraged women's public participation and voice. By including her antislavery poem "A Word to the Weary" in *Lichen Tufts* alongside her nature writing, Wright suggests the intersections of environmentalism and social justice and encourages her readers to persist in their consciousness and advocacy for both.

While Wright regards the natural world as universally liberating, she highlights the particular importance of women turning to nature for a sense of freedom and empowerment. After describing in "Into the Woods" the ways in which her own camping party discarded the rules for the genders (for instance with Alfred professor and future president Jonathan Allen cooking for the group),

4 *Proceedings of the American Anti-Slavery Society, at Its Second Decade, Held in the City of Philadelphia, Dec.3d, 4th, and 5th, 1853* (New York: American A. S. Society, 1854), 36.

Wright weaves the link between environmentalism and women's equality throughout the remaining essays of *Lichen Tufts*. She addresses deficiencies in women's education and restrictive fashion codes and urges her readers to adopt the "nature cure" for mind and body as an antidote to harmful gender norms. An activist involved in all of the major progressive reform movements of her day, Wright was devoted to universal enfranchisement for all men and women. In 1869, in the midst of a major rift in the US suffrage movement, Wright's role in a couple of pivotal events reveals her stance as a feminist and equal rights activist. Social reformers who fought for women's suffrage were divided over the Fifteenth Amendment, which granted voting rights to all men regardless of race but excluded women. Wright was very likely the "Elizabeth Wright" listed as the representative from Texas (where she was living for a brief time) at a May 1869 convention of the American Equal Rights Association, a group focused on universal equality and suffrage for African Americans and women.[5] At this convention, Elizabeth Cady Stanton and Susan B. Anthony announced that they were splitting from the organization and from the campaign for universal suffrage. Stanton, Anthony, and like-minded feminists withheld their support for the Fifteenth Amendment, because they objected to the prioritization of voting rights for African American men before women. They formed an organization devoted strictly to women's suffrage, the National Woman's Suffrage Association (NWSA); meanwhile, the activists on the other side of that rift formed a separate women's suffrage organization, the American Woman's Suffrage Association (AWSA), which supported the Fifteenth Amendment and strategically fought for women's suffrage amendments at the state level. The AWSA's organizational convention, held in Cleveland on November 24–25, 1869, listed Elizabeth C. Wright as the vice president from Texas. It is highly likely, given records of Wright's move from Paris, Texas, to Kansas in November 1869, that this is the same Elizabeth C. Wright who wrote *Lichen Tufts*.[6] Given her radical voice as an abolitionist, it makes sense that Wright aligned herself with the organization that focused on intersecting forms of oppression and fought for the rights and enfranchisement of all people.

While actively contributing to social justice campaigns, Wright also continued her work as a natural scientist. Her *Putnam's Magazine* (October 1869) article, "Something About Fungi," highlights the value and methodology of

5 "American Equal Rights Association," *New-York Tribune*, 13 May 13, 1869, 5.
6 The likelihood that the Elizabeth Wright listed at the first AWSA convention is the same Wright who wrote *Lichen Tufts* is also noted by Jessica Brannon-Wranosky, *Southern Promise and Necessity: Texas, Regional Identity, and the National Woman Suffrage Movement, 1868–1920* (PhD diss., University of North Texas, 2020), 36–37, note 27.

her main area of expertise, mycology, a branch of botany that she felt was especially important given the plentiful native fungi species in the United States. In
November 1869, after her brief time living in Texas, Wright moved to Kansas,
where she continued her work as a botanist and teacher while starting a new
chapter as a homesteader. She laid claim to an eighty-acre plot in Marshall
County, Kansas, and, in keeping with the environmental ethic she outlined in
Lichen Tufts, she immersed herself in the local natural environment of her new
home region. Findings from Wright's field research appeared in major pioneer-
ing catalogs of native plant species in Kansas in the 1870s. Around this time,
Wright also recruited Lyman Jewell, a fellow teacher (and former student)
from Keating to join her in Kansas. Jewell claimed the eighty-acre lot next to
Wright's, and they were married in February 1876. Characteristically defying
social rules for women, Wright married late in life to a man twenty years her
junior, and it appears that her marriage to Jewell may have been polyamorous.[7]

Wright's openness to an unconventional marriage could reflect some of the
subversive ideas of the Free Love movement, which promoted freely chosen
love as an alternative to the legal and economic bonds that defined traditional
marriage. There is no evidence to suggest that Wright was an adherent to the
Free Love movement, but it was a popular set of ideas in the Burned Over
District and may have influenced Wright's generally open-minded attitude
about her marriage. Given this movement's intersections with other areas of
progressive culture that can be traced in Wright's life and writing, it bears con-
sideration as a relevant context. With its focus on empowering people to make
decisions freely about sexual relationships, family planning, and birth control,
Free Love was regarded by many feminists of the day as a path to women's liberation from patriarchal marriage laws. The movement's most famous spokes-
person was Victoria Woodhull, best known as the first female candidate for
US president (nominated by the Equal Rights Party in 1872). In addition to
being a radical feminist and Free Love advocate, Woodhull also practiced
Spiritualism and belonged to the largest Spiritualist community in the world,
Lily Dale, New York. Spiritualism, founded in western New York in 1848, was
based on the belief in an afterlife and the ability of the dead (spirits) to com-
municate with the living, usually through a medium and during séances. More
than a set of religious beliefs, Spiritualism reflected the Burned Over District's
radical culture and progressive values. The movement appealed to social justice
activists, and it empowered women as leaders and mediums. Lily Dale, which

---

7   Details about Wright and Jewell's marriage are included in Jewell's journals, which are
    transcribed and contained in a binder of family diaries and papers, "The History of the
    Lyman Jewell Family," Marshall County Historical Society, Marysville, KS.

still exists today as a thriving community of practicing spiritualists, hosted suffrage meetings and lecturers, including such prominent suffragists as Susan B. Anthony, Anna Shaw, and Carrie Chapman Catt. Radical young Quakers were especially drawn to the movement, as it combined religious devotion with social reform activism and comported with their progressive ideas about women's roles in society. All of this said, it's not surprising that Elizabeth C. Wright and her husband, Lyman Jewell, were receptive to the practices of Spiritualism and participated in séances, nor that they had an unconventional marriage.[8]

Lichen Tufts is brimming with the radical philosophical, social, and religious experiments from the New York-Pennsylvania border region of Wright's childhood, education, and early professional life. At the same time, Wright's nature writing and philosophical reflections are connected to a national intellectual tradition. In shining a light on the beauty of her beloved Allegheny wilderness, Wright draws upon the American transcendentalists' passion for environmental consciousness, liberating sense of deliberate living, and spiritual communion with the natural world. Explicitly aligning her own nature treatise to Henry David Thoreau's, in "Into the Woods," Wright recalls carrying along her copy of *Walden* and reading passages aloud to her camping party. As Daniel Patterson has noted, it is "undoubtedly the first time that a fellow nature writer pays tribute to the project of cultural change that shapes Walden."[9] In the essays of *Lichen Tufts*, Wright outlines an environmental ethic that highlights the far-reaching benefits of an intimate acquaintance with the natural world, especially in one's home region. She makes a compelling, transcendentalist case for the liberating effects of nature on a person's character and philosophical outlook in "The Nature Cure—for the Mind": "As soon as you begin to enter into the arcana of nature, you feel the shackles of outward customs grow loose, and the liveries of many servitudes drop off, as a bird moults its feathers. Nothing is done suddenly, for Nature has plenty of time—all the time there is—and is never in a hurry" (91). With echoes of Emerson's philosophy of self-reliance, Wright proposes that such an immersive encounter with the natural world is the key to an elevated consciousness, unburdened by the prejudices and opinions of others and freed from the restraints of social and intellectual conformity: "Other men's opinions of you are now of less import to you than your own. You respect yourself—you approve yourself, therefore you can afford to be censured

8 Elizabeth (Wright) and Lyman Jewells' participation in Spiritualism and séances is described in "The History of the Lyman Jewell Family," 67.

9 Daniel Patterson, "'I commend you to Allegany underbrush': The Subversive Place-made Self in Elizabeth C. Wright's Treatise on Nature, *Lichen Tufts*," *Legacy* 17, no. 1 (2000): 33.

for what is best and bravest in you—you can afford to be misunderstood" (92). For Wright, the risks and rewards of nonconformity were particularly high for women of her era. Wright spends much of *Lichen Tufts* outlining both the bleak consequences of gender codes on women's health and education as well as the particular value in women turning to the natural world for liberation. Wright's feminist vision for environmental consciousness as a path to resisting the patriarchy especially aligns with transcendentalist author Margaret Fuller. Based on her journey to the Great Lakes and Niagara Falls, Fuller's travelogue *Summer on the Lakes, in 1843* is an important precursor to Wright's *Lichen Tufts*, as it features early ecofeminist commentary on the impact of rapid industrialization in a patriarchal society.

Wright's work, especially her unapologetically radical abolitionist stance, is also reminiscent of the prominent abolitionist and author Lydia Maria Child. A beloved novelist and children's author, Child's career suffered a blow in 1833, when she published the book *An Appeal in Favor of that Class of Americans Called Africans*, in which she argued for the immediate emancipation and enfranchisement of enslaved people. Instead of being deterred by the severe condemnation of her abolitionist writing, Child became a leader in the movement, serving as editor of the weekly newspaper the *National Anti-Slavery Standard*. Much like Child, Wright fearlessly broke society's codes for women, publicly expressing her radical views and fighting for the rights of the oppressed. Indeed, as much as Wright is exceptional for defying restrictive, gender-based norms and rules, she is also representative of the countless women of her era who refused to be silenced or excluded from the public sphere. Wright's voice is a reminder that, long before they were granted the right to vote, women's voices were indispensable to public policy and social justice reform.

In her work as a natural scientist, like many women occupying spaces traditionally reserved for men, Wright did the double duty of contributing to her field while also advocating for reforms that would enable more women to participate in them. While it may seem like a digression from her camping narrative, her lengthy critique of sentimental "language of flowers" books in "Into the Woods" exposes the popular books that were marketed to girls and often took the place of rigorous and valid botany reading. Wright was especially concerned about the typically paltry science education received by girls because she believed a deep knowledge of the natural world is vital for self-elevation and liberation and, in turn, beneficial to society as a whole. This and other tenets of Wright's environmentalism are the most enduring and prescient aspects of her life's work and philosophy, while one aspect of her nature treatise was outdated almost immediately upon its publication. As Patterson points out, Darwin's *On the Origin of Species* was published in 1859, just before *Lichen Tufts* and before the Darwinian theory of evolution was well known and established in the science world and beyond. As a result, Wright's theory of nature reflects the widely held creationist understanding of the natural world, that plant and animal species were fixed by God. But, as Patterson makes clear, that belief is just one part of a complex and forward-minded theory of nature: "She spoke during the last days of special creationism, yet her textual representation of nature clearly reveals that she was moving toward an ecological understanding of the nonhuman environment."[10] Wright's nature treatise is a passionate argument for the far-reaching benefits of an immersive, conscious relationship with nature. *Lichen Tufts* is a plea on behalf of the natural world as well as for humanity, as Wright believed in the reciprocal rewards of environmental stewardship. With the modern-day turn to the natural world as a site of concern and advocacy—as well as a site for personal lifestyle elevation and even aesthetic inspiration (the growing popularity of the Norwegian concept *friluftsliv* comes to mind)—the wisdom of her ideas about the "nature cure" and the urgency of environmental awareness are remarkably relevant today. Over a century and a half since the publication of her prescient writing, what would Wright say to a reader today seeking physical and mental wellness, mindfulness, and a respite from the pressures and distractions of society? Her advice is simple and as timely as ever: "'Go to Grass.' It will do you good."

10 Patterson, "'I commend you,'" 3.

LICHEN TUFTS.

INTO THE WOODS.

WE were tired and wanted a holiday, so we went off into the woods, out of the way of finery and etiquette, and conventional rubbish, where we should escape from fashionable twaddle, gossips, and flirts—from humbugs and household botheration, and be free to rest and refresh ourselves at leisure. Such an elimination of the ordinary burden of life's occupations would not only leave us free, but make us rich with an unheard of wealth of hours, per diem, at our own disposal.

So running the gauntlet of all manner of croaking prophecies about colds and rheumatisms, and spiders crawling into our ears, and caterpillars creeping in at our mouths, and of all manner of shocked proprieties wagging dissenting heads at us, and a prospective storm of "I told you so's!" to hail our disappointed and untimely return, we packed up our bed and board in the narrowest possible compass, and went off, like overgrown children, to play in the woods.

This simple performance, like many another, undesignedly put a variety of pretty professions to an unexpected test of their genuineness. It was surprising to find how many readers of verse, who professed to love poetry, and to appreciate the enthusiasm of the poet-lovers of Nature, shrank from any actual participation in a poetic life. Mr. A., who could quote volumes of poems, thought a man who could live comfortably at home, and have a good hot dinner every day, would be a fool to go into the woods where hard beds and cold dinners would be inevitable.

Then there was gallant Mr. B., who compares women with angels, and grows ecstatic over a fine voice or a graceful carriage, was afraid to trust himself in a camping out excursion "burdened with so many incumbrances!" meaning *ladies*.

Mrs. C., who is fond of Shakspeare, maugre his outlandish heroines in boys' clothes, was shocked at women exposing themselves so, and doing such unladylike things as the ladies of our company proposed to do; and sentimental

1

Miss D. thought it would injure her complexion, and though she could admire
Ellen Douglass,

> "Though the sun, with ardent frown,
> Had slightly tinged her cheek with brown;"

still she feared tan more than she loved sunshine and poetic living.

Mr. E. told eerie stories about belated hunters unwittingly camping on a
rattlesnake's den, or at the foot of a panther-haunted tree, or in the favorite
retreat of a family of bears, less amiable than those which entertained little
Silver Hair in their domicile.

Mrs. F. imagined that there was a tornado somewhere, keeping on purpose
for us, when we should once get inextricably into the forest, when the tall trees
could come crashing down upon us from every hand; and the G.'s dreaded wild
fire and wilder flood, lost bridges, unfordable rivers, and nameless disasters.
One would think, if they were to be believed, that our quiet and "grand old
woods" were huge receptacles of dangers and discomforts, unconquerable and
terrific. But some of us were good woodsmen and knew better, and longed for
the cool pure liberty of their hidden depths.

We were wearied with the experimental rehearsal of life's drama, and ready
to go back of all rehearsals and acting, into the forests and grottoes where
the air breathes poetry, and all the elements of grander dramas than ever we
have enacted, are created and exhaled by rock, and tree, and moss—by cool
spring and shady river—by many-toned birds, and bright-hued insects, and
shy wild beasts—by fog, and cloud, and wind, sunshine, and rain, and dew. We
had a mind to lie out under the skies and catch any divine ideas that might
fall, with falling stars, on the soul not shut in from them by lath and shingles.
We would lie on the bosom of Mother Earth and listen to her breathing, and
thereby interpret her dreams. We would hush still the life that was in us, and
listen for "the sound of growing things." Perhaps if our own hearts would beat
silently awhile, we might be able to hear the pulses of the green-blooded plants,
and the breathing of leafy lungs. If one of us should be endowed with genius
enough to write out a faint transcript of the divine poem we found growing
wild in the wilderness, our croaking neighbors would perhaps shed tears of
sentimental rapture over the beauty of the fragmentary transcript (especially if
they did not know who wrote it), although they believed the living poem was
not worth taking the trouble of going to see. It was all the pleasanter, however,
that they preferred hot coffee and feather beds to cold water and hemlock
brush, for we were not in any danger of being crowded or jostled in the Temple
of Sylvanus, to which we were going, as devout a caravan of pilgrims as ever
visited any shrine.

If we could only have gone as the birds go, unburdened with baggage and unmindful of raiment, we should have been superlatively happy. It was a very great relief, even with these drawbacks, to go where we might wear the same apparel day after day, without remark or change, and this apparel too of the simplest and most convenient character. It would be as incongruous as stupid to carry finery into our democratic woods, where hemlock knots and bramble brush are no respecters of persons, and will tear a dandy's rigging or a fine lady's flounces as placidly as they rend a beggar's rags. I commend you to Allegany underbrush, ye who hate frippery and fandangles, and have a liking for seeing them put to the proof by contact with what is genuine, and strong, and beautiful. But "they that wear soft, raiment are in king's houses," or at least far enough from these sombre retreats of verdure; and as the beggars, too, are only to be found in the neighborhood of fine houses, we enjoyed a respite from the heavy presence of both the livers for flattery and the livers on broken victuals.

It is bad enough any time to have some lazy mendicant thrust a lying document under your nose, and assure you in good fair type that the holder is some Italian patriot, whose family are waiting in the interior of Mount Vesuvius for him to collect the means of having them dug out, and that as he can't speak English he has to get his begging done by his *printed* paper, which we all know is but one of a whole edition of such documents carried about by similar vagabonds. Ten to one if you tell the scamp he can understand English, he will swear to you in a most rascally accent, but quite comprehensible, that he really can't talk at all. But these poor liars "of the baser sort" are not near so offensive as many respectable liars we meet every day, and whose polite falsehoods we dare not kick out of doors so heartily. When your soul is utterly weary with shaking hands with pretence, and conversing with make-believes, you too will be ready for such a plunge into the wilderness.

A few requisite qualifications were needed, and a few equipments necessary for our outfit, and though they were few, they were rare enough to make our party none of the largest. The members of it must be able to stand fire and water, and be of sterner stuff than dolls or carpet knights are made of. It would not do to have one of us get frightened at a bear track or be uneasy at sleeping in a wall-less lodge in the wilderness. A love of Nature and Adventure, and an indifference to Luxury, were requisite. These three prime requisites for an explorer or a hunter, could not be dispensed with, even for the Lilliputian undertaking we had in hand.

When equipped we looked a singular group of animals enough. Our captain wore a scarlet upper garment, called by courtesy a shooting jacket, but which to the uninitiated bore a decided family resemblance to a red flannel shirt, fastened by a leather belt buckled around his waist, and when this was surmounted by his shot pouch and powder horn, he looked decidedly picturesque, and would have been gobbled after by all the turkeys in town, if we had been in town. The ladies were metamorphosed by short dresses, broad hats, and thick shoes, into as many substantial wood nymphs, ready for scaling rocks and fording streams.

We left the railroad at Great Valley, for the woods and river here are still in possession of the aboriginal inhabitants, the grave and friendly Senecas.

There was an agreeable contrast between the manners of the white inhabitants around the dépôt, and the red inhabitants round about. The Yankee curiosity of the former was all agog to know who we were, and where we came from, and what we came for, and what could possess us to do so outlandish a thing. They reminded one of the comments of the poet Wordsworth's neighbors, who believed his visitors to be a doubtful and suspicious set of persons, because he and they went "out o' nights" to enjoy moonlight views.

The Indians, however, to whom camping out was a more natural phenomenon, and who were not plagued with so great a desire to meddle with other people's concerns, took it in a very matter-of-course sort of fashion, and did not take the trouble to stare after us, nor to make impertinent inquiries. The courtliest politeness could not have ignored the singularity of our appearance and proceedings more completely. Much of the picturesqueness of their character has been civilized away, but the serious dignity and hospitable courtesy remain.

Our first camp was unfortunately chosen, but we comforted ourselves with the doubtful adage, that "A bad beginning makes a good ending," and made the best of it we could. Our Izaak Waltons had a trout brook down in the programme, and inquired for one at the dépôt, and a young Seneca accordingly ferried us over the river, and piloted us a quarter of a mile through the woods, to a deeply shaded, mossy bank, overlooking the desired trout fishery. We soon found, however, that our location was possessed of several qualifications not down on the programme. The thick damp woods swarmed with mosquitoes, which invaded even the dense curtain of smoke we hung to the windward of our hastily erected tent. We were tired and sleepy, and having spread a pallet of hemlock boughs we lay down to rest; but our visitors were hungry, and thirsty, and importunate, and had greatly the advantage of us in numbers, and so made our first essay at camping out rather more spirited and musical than tranquil or agreeable.

We arose in the morning devoutly thankful that our ease-loving companions had stayed at home, and sincerely glad that our camp was easily moved to some more eligible spot of earth. It was not so easily done, however, but that it might have been a great deal easier if we had not encumbered ourselves with so much luggage. We had agreed at the outset that we would take as little as would suffice for our needs during the expedition; but all being novices at this kind of life, and some of us having such hospitable ideas of the sufficient, we found ourselves when all assembled, entrenched behind a rampart of goods and chattels—provisions, blankets, artillery, and ammunition—books, portfolios, and spare garments, so that we felt as "hard set" to know what to do with our supplies as did the man who drew an elephant in a lottery. The comforts of life are very inconvenient travelling companions. How were we to transport our civilized rubbish through the woods? If we had been Dutch peasant women, we could have carried it on our heads, but as our crania had been accustomed to bear only less material burdens, we had to cast about us for some different mode of conveyance. I made "Walden" play Balaam for us, and "curse us" all that aggregation of the plagues of luxury our own "cursing" was unequal to.

He says: "The more you have of such things the poorer you are. Each load looks as if it contained the contents of a dozen shanties; and if one shanty is poor, this is a dozen times as poor. Pray, for what do we *move* ever but to get rid of our furniture, our *exuvice;* at last to go from this world to another newly furnished, and leave this to be burned. It is the same as if all these traps were buckled to a man's belt, and he could not move over the rough country where our lines are cast without dragging them,—dragging his trap."—"I think that the man is at a dead set who has got through a knot-hole or gateway where his sledge load of furniture cannot follow him."—"It would surpass the powers of a well man nowadays to take up his bed and walk, and I should certainly advise a sick one to lay down his bed and run. When I have met an immigrant tottering under a bundle which contained his all—looking like an enormous wen which had grown out of the nape of his neck—I have pitied him, not because that was his all, but because he had all *that* to carry. If I have got to drag my trap I will take care that it be a light one and do not nip me in a vital part. But perchance it would be wisest never to put one's paw into it."

When we crossed the river to encamp by our trout brook and mosquito factory, we had passed through one of those "knot-holes" where our baggage had to remain behind at the hotel. Now we returned to take up our burdens again and move on.

It was a glorious July day, blue and golden, with the fiery languor of summer's noon, quivering in the heated air, only stirred now and then by a cool breeze winding up the river, like a pure and fresh aspiration in a life of indolence and passion. All day long the active portion of our party had hunted and fished without catching anything, or had romped in the woods and on the river without having the fear of torn garments or wet feet before their eyes. It was utterly delightful to let ourselves loose, and live freely; to have no rules for coming in or going out, for rising up or sitting down; to be emancipated from the bondage of the ceremonial law, and do what pleased us best was paradisiacal enough. The girls tried to learn to throw stones, but did not succeed very well. When the unsuccessful hunters returned, Elvira took a lesson in shooting at a mark. It was awkward business, though she finally succeeded in taking aim, but somehow or other her fingers did not avail to make the thing go off. Two or three attempts failed, and she was on the eve of giving up in despair, when a mosquito settled comfortably on her outstretched finger, and began to try the flavor of the savory fluids hid under that thin white cuticle.

Elvira is very sensitive on the subject of mosquitoes. She has a great tenderness towards them, and they make strong and lasting impressions upon her. Owing to this amiable weakness she is agitated and alarmed at their approach, and this unexpected salute so startled her that the bitten finger closed convulsively, and the gun was discharged. No doubt the load hit something, for it is not probable that it is yet wandering unresting through the air, but *what* it hit is unknown. Elvira insisted that she fired at the mosquito, and killed it—I, for one, believe her.

Having thus fired our sunset gun, we emigrated from our camp, and pursued our way. Our passage had been secured in a scow, bound for the mouth of the Kenjua, and loaded with provisions. As it was not quite loaded we were obliged to await its completion at the hotel. There was a mingling of odors about the mansion, common about kindred places, and no more disagreeable than at any other time, I suppose; but coming in from the fresh fragrance of the woods, the smells of tobacco and whiskey, onions and pork, soapsuds, codfish, coffee and supper, all united, were disgusting to the last degree, especially to those who could not "eat in faith" in any tavern, because too familiar with the kitchen.

They were a long time loading the boat, and when it was done, they came and told us that one of the hands was too ill to proceed, and that they should not start that night. We held a brief council to discuss the question, "Which shall we do, stay in the inn, or go out of town and encamp?" The "town" was a little pocket affair, easily got out of, but it was now sunset. But then July twilights linger longer than most welcome guests do, so that much could be done in that pleasantest hour of the pleasantest time in all the year. Susie seemed to have a mind to try the benefit of a feather bed, but where was the use of coming sixty miles from our feather beds at home, to sleep in a tavern where all manner of people sleep? For myself I felt with a selfish emphasis, that all the sleep in the world was out of doors, and could be found nowhere else that night.

I had grown up in woods like these, and they were home to me. I had been absent from them two years in the west, and had longed with more unspeakable homesickness for the evergreen woods and mountain air, than for home or friends. Among them again I could not afford to waste an atom of their riches, so longed for and coveted while I was gone. The mountain echoes were ravishing music to ears wearied with the flat echoless sounds of prairie land. Who could afford to sleep indoors and thus lose any part of that grand Oratorio of unwritten music played by the wind on a wilderness of harps?

There was a whizzing of car wheels, and a rumbling of wagons—a clatter of tongues, and the ominous scraping of a fiddle on the dépöt steps opposite the inn. There were several young ladies with many-tailed head-dresses and frightfully big hoops, who were menacingly polite. These were more to be dreaded than a return of the ague. The noises jarred harshly on souls longing for sweet accords and divine harmonies.

A few rods off on either hand, stretching far beyond our sight, lay the cool, dim forest with its music and its silence whispering together in the twilight. When out of this present little sphere of racket, one could hear the solemn anthem of the far off pines, and the monotonous rush of living waters. Away from this poisoned air, beyond this clang and discord, yet close at hand, a wilderness full of fragrance and music waited for us.

The old boatman, who was an old hunter too, took our baggage and some of us in a skiff, half a mile down the river, and landed us in as beautiful a spot as we could hope to find. Our grizzly and half-tipsy Charon appreciated our errand heartily. He had no wonderment about the oddity of our choice of amusement as our civilized fellow citizens had had. *They* could hardly be convinced that we really meant to do so very odd and uncomfortable a thing as to go out and try to live the poetry which was only fit to read, but the Indians took it as a matter of course, and our boatman delighted in it. He had lived in the woods for many and many a year, he said, and had followed the river until he could pilot it in the dark. We must certainly come to his place on the Kenjua. It was the greatest kind of a place for trout. *Kenjua* was Seneca for *fish creek*, and it was rightly named. He would delight to show us everything, rattlesnakes and all.

We landed on the pebbly beach under a spreading sycamore, which leaned its heavy head waterward, and seemed to be looking at itself in that ever wavering mirror, as it reflected its gleaming white arms and made them sway and beckon upwards, though the tree itself stood still. But the deep water shoaled to a ripple here, and each separate facet of every wave set up for a mirror on its own account, reflecting what it could at a miscellaneous variety of angles of reflection and refraction, so that the sycamore had the pleasure of beholding its own clearly defined and pointed foliage run together into an indefinite mop of green, dancing incessantly on the changeful surface. There was a curve in the river here like a silver bow, and on the opposite side, which was the outer rim of the crescent, the water was tranquil as a pond. From its waveless edge up to the region of clouds, rose the hill abrupt and dark, clothed to its very summit with an unbroken mass of evergreens, rising tier on tier up the steep side of this "mountain wall," whose shadow lay, like a fragment of midnight, on the pool below. We encamped on the grass under the lighter foliage of young trees which clothed our side of the stream. A grassy open space spread towards the river before us, and a sheltering thicket gloomed behind. We built three bright driftwood fires in a triangle, and within the area we spread our blankets in gipsy-like groups, and with a roof of sky and stars above and walls of green tapestry about us, we lay down, safe and happy, and watched the sparks fly up like showers of stars among the leaves, and saw the smoke go rolling upwards like home-brewed clouds, going to seek their kindred above. A grateful content, a peaceful rest, such as comes to happy children, settled upon us like dew upon the grass, and those who did not sleep lay listening to the "voices of the night."

He who lives under the eaves of the forest will learn the voices of the trees. They have a definite speech which the initiated can understand. Here lies the germ of truth which makes the charm of the Oriental fable, that some persons can learn the languages of beasts and birds and all living things. Even our dull Western ears can learn, by long listening, to distinguish the differences in the whisperings of trees. If you sit at a street window in one of the thoroughfares of a cosmopolitan city, and listen to the varied gibberish jabbered in the streets, you will soon learn to tell the Irish and Scotch brogues apart, and as easily tell the Swedish from German, or Spanish from Italian or French, without knowing any language but your mother tongue. So in the woods. You may not know what the leaves say in their musical whisperings, but you will know their several tongues. The voices of trees are as characteristic as the voices of the human race. The poet eye of the traveller was in love with the tall tropic palms of the Indian isles, and he thought it would be life enough for a poet to lie ever under that broad, plumy foliage, and be fanned by those leafy wings. But the sea breeze came and whispered, and the harsh, reedy hissing of the palm-tree's greeting disenchanted the finely-attuned ear, and the poet soul returned to its allegiance to the pines of his native hills. We should not have believed him to be a poet if he had not remained loyal to the oldest of the forest kings, and greatest minstrel and bard of them all. The founder of the cone-bearing dynasty lies buried a thousand "aeons" deep under the coal-measures, on the spot where it held up its green æolian harps to catch the primal winds, on which no wings had ever been spread—which had never thrilled to a morning bird's song, nor vibrated to the notes of a grasshopper. The empty sea beat on the desolate shore, and only the lonely shell, and this old Homer of all the trees, heard its throbbing and answered its song. That land has sunk down in the sea, and the waves have rolled over it since then. That ocean has frozen to glaciers and melted again. The land has risen out of its long bath, and been peopled by strange creatures, and that sea has teemed with monstrous shapes, but the cone-trees from the mountains have answered the song of the waves through all the ages and their changes.

Older than man, or beast, or bird—older than the crumbling soil we tread, that has been deposited here, particle by particle, for thousands of years—it stands, sombre and tall, over the deep-buried graves of its primeval kindred, and holds up its æolian harp-strings to the touch of its immortal lover, the wind. Together they chant the heroic song of the Ages, at once a requiem and a battle-cry—a prophecy and a dirge. There is in it a reminiscence of the song of the morning stars when they sang together—a lingering echo of the great shout of joy over the new creation, uttered by the sons of God; and the dying sigh of the dead Ages and their lost types of being, is repeated in the song of the Pine.

A stranger would hardly know the roar of the wind among the hemlocks from the deeper and hollower sound it brings out of the longer leaved tops of the pines, but a novice could distinguish the difference in their whispers. It is the difference between a robin's "Good morning!" and the wood-thrush's "Good night!"

There are two kinds of aspen trees here in the underwood, distinguishable from all other trees by the fairy tap, tap, tapping of their leaves against each other when the other woods are still. I know no common names by which to distinguish one from another. They are both known hereabouts by the names of "Popple," "Quaking Asp," and "Aspen." The *tremulöides* is the-more restless and sensitive of the two, as well as the more graceful. Its smooth, heart-shaped leaves dance on their long flattened stems long after the wind has gone by, and by the constant beckoning of its many hands seems calling the rover back again. A sound like the clapping of elfin hands issues from it, and seems to be calling the invisible air with an oft-reiterated "Come! Come! Come!" The more sleepy *grandidentata*, which rolls its fuzzy leaves out of its white buds a fortnight later in the spring than its early rising relative, is a slightly graver personage, and does its tap-tapping to a slower tune. Its leaves dance, but in a more solemn, or mayhap a lazier measure, for it is not always easy to tell laziness and drawl from gravity and solemnity. The grandiose name of the larger aspen, however, impresses one with the notion that the rapidity of the movements of the *tremulöides* would be beneath its dignity.

The magnificent magnolia acuminata of the Alleganies, known by the stupid name of cucumber tree, shudders audibly when struck by the wind. It is a grand and peculiar tree, well nigh as conspicuous in its individuality as the pine. Taller than any neighboring deciduous tree, it lifts its top straight above its compeers, and even its branches and lesser boughs follow the proud aspirations of the trunk, and lift themselves, massive and angular, right upward at the ends. There is no pliant swaying, no graceful drooping of outermost twigs or pendant foliage, but all are as upright as a Prussian regiment or a Puritan Sunday school. It is against its nature to bend, and the hoarse shuddering sigh with which it meets and buffets the blast, is uttered by no other tree. Its great oval, spice-breathing leaves carry off the wind in spouts, like a multitude of eaves troughs slanting to the leeward, and the rushing of these divided currents has the disturbed and warlike tone of opposing forces. It affects one like a remembered trumpet blast.

Then there is the birch. It grows to a great tree here on the Alleganies, and has an elm-like foliage, and an odor of winter-green. I cannot describe its voice, for it is to me so associated with a very different sound, that I cannot think of them separately, though they are not alike in the least. In our old sugar bush there used to be a great black birch which had many years agone taken root on a fallen log, and its long roots had run down on either side to the earth and taken fast hold there while the fallen tree decayed away from beneath it, leaving our birch standing on a five-legged stool of its own twisted roots, in the air. Another fallen log lay near by, covered with a thick mat of yellow, featherlike mosses, and on this used to stand a patriarch grouse, or "pheasant," or "partridge," as he was called, and wake the dreamy echoes with his drumming. We very rarely saw him, for he was a shy bird, but we heard him many times a day in his season, and found his tracks there. We children used to tap the old birch and catch its profuse sweetish sap in a little trough, to drink of its diluted spicery; and that draught and its neighbourhood, and the tremulous thunder of the grouse's wings, with the hum of the wild bees which used to drink at the same place, are all called up by the birch's voice, and become part and parcel of it, past the power of analysis to separate them.

No more can I remember the voice of the maple without the crackling of the fire and the singing of the heating sap-kettles, and the tinkling drip of the drops from the trees into the receivers all being mingled with it. Nor has memory any language for the sycamore unassociated with the dash of rippling waters. It is not to be regretted, however, for such associations make much of the poetry of the unwritten language of trees. Would you wish to forget in connexion with the long drawn "Hush-sh-sh!" of the willows by the stream, the plunging sound of the musk-rat as it leaped from its perch on shore and swam to a place of obscurer safety? Or could you afford to drop out the sound of the occasional leaping of a fish from its element, to seize a passing fly; or the creaking whiz of a dragon-fly's wings, as it darted to and fro. Would you like to remember the beech or oak without the squirrel with puffed cheeks, crying "chip! chip! chirrrrr!" at you, as you tried to measure the measureless "O o o o o o o o!" sung by the wind in the tree tops?

You might as well try to think of a sleeping village without its barking watch-dogs guarding it, or remember

"Yon ivy-mantled tower,"

without thinking of its moping owl, "complaining to the moon."

There is a language too, which is not Spoken by the voice, but uttered in look, or motion, or taste, or odor. The resinous smell of the coniferæ speaks as distinctly as their wind-swept leaves. The penetrating, disease-dispelling

odors of tar and turpentine, rosin and balsam, have the same primeval reminiscences as the voices of the trees that produce them. They suggest questions
concerning the probability of some post-alluvian beauty decorating herself
with beads of fossil rosin, as the belles of our day wear ornaments of amber.
They may guess of the rosin, as we guess of the amber, how it was distilled in
the laboratory of the sea-loving cone tree's trunk in some unknown ante-hu-
man era. The bitter spice of the magnolia and tulip trees tastes and smells so
suggestively, as though they had had human experiences, that we wonder if
those trees never travelled, nor loved, nor hated, nor suffered, nor enjoyed
like us.

When we read of birch and willow twigs being found undigested in the
stomach of a fossil mammoth, we wonder if some birch-eating school-boy's
fossil remains might not suggest to the post-alluvians, the notion that the
genus boy was birchivorous in its habits. We alluvians would be more apt to
class the pedagogue there, but geology would hardly teach so much to the scientific of the next era.

There is an odor of antiquity about the birch and such trees as are found
as fossils, though it is but a mediæval sort of antiquity after all; its ancient-
ness toned down and mellowed by the far older and darker background of the
cone-bearing ages that came before. This sort of reflection and research is the
preparatory course of study needed in order to understand the language of
trees. *These* dead languages are superclassical, and have the advantage of being
real knowledge and living poetry, beside the discipline and dead lumber which
are the recommendations of the mere dead *human* languages.

Within the sphere of purely human oldness there are some trees and herbs
geologically both ancient and modern, which are venerably hoary with the rime
of the winters of many generations of men. The cedars of Lebanon would seem
as old as Lebanon itself to our thoughts, and as venerable as the temple of
Solomon, and as mysterious as Masonic rites, even if we knew nothing of the
story of its fossil kindred. And to those who, as children, took the oar-like seeds
of the ash, and hung them to the lip of the cypripedium, the "whippoorwill's
shoe" of the Senecas, and called them the oars of a fairy-boat, the ash becomes
an ancient and legendary tree, even before they come in after years to read the
story of the evergreen ash tree yggdrasill, whose roots the dragon Niddhögg
gnawed, and whose branches sheltered the Norns. Those fairy oars have rowed
that legend through the age of fable down to the matter-of-fact sea of modern
literature, and yet not a vein of that charmed tissue is broken, and the sharp
lamina of the blade is as delicately veined and nerved as in the days when the
Eddas were new. The mystic ceremonies of the Druids stick to the viscid berries
of the sacred mistletoe, and something of the smoke of their altars lingers in the
scent of the oak on which it grew.

But these legendary reminiscences are only historical, and not the natural language of the trees themselves. They speak of them only to those who know the story, but never tell the stories themselves even to the most intimate of their acquaintances. They are no part of the tree history proper, only an episode in which there is more of the history of men than of trees, and we cannot expect trees to learn a new dialect to add to their speech, that they may be able to prate of men. Yet we, being human, like this odor of old human ideas clinging to them, and would willingly interpret their speech awry in order to find our fore-fellows mentioned in their chronicles.

The less pretentious herbs of the field and forest have a language as pithy and vigorous, according to their kind, as the utterances of the lordlier trees, but it is no such "language" as the plants are accused of in your "Flora's Lexicons" and "Flora's Interpreters."

Imagine Flora treading on the margin of vanishing winter snows, leaving flowers behind her for footprints, and beckoning the leaves out of their buds, and the flowers from their occult hiding-places, with her floating garments and drooping garlands, wearying her wreathed arms by lugging around an eight-cornered dictionary, and attended by an ominous interpreter tagging at her heels! Bah!

This sentimental flummery could no more be mistaken for the "language of flowers" by anybody who had ever heard the real language, than the quizzing schoolboy's "Iggery, wiggery, foggery, youggery!" could be mistaken for classic Latin by a Cambridge Professor.

But it is surprising how many people take such stuff for poetry. It seems to be taken for granted that everybody who likes poetry or flowers must be maudlin enough to like these satiny, perfumed, illustrated books, with a flower's name in two or three languages, (and not always correct at that) at the top of the page, and selected fragments of the productions of poets and poetasters strewed down the space below the name and motto.

"We even find a few pages at the end of our Botanical text books devoted to this twaddle, and when the books have been used awhile we find these pages have been used more than any other, the leaves falling apart quite naturally at the silliest places.

One of these vapid prettinesses, designed for sentimental young girls to read, is entitled the "Floral Diadem," though poor Flora must be as mad as King Lear before she will wear such a ridiculous diadem over her "ambrosial locks." This book, like its congeners, makes some botanical professions. One of the first articles in it contains the following erudite morsels. "It (the Spring Beauty) appears in the verdant meadows, upon the grassy hills, and in the shady groves, lifting its pink striped blossoms above the ground. *It varies in color from sky blue* (!!) *to a fair white* (!!). But by far the most common and the most beautiful are the variegated ones, delicately striped with pink and white. It is one of the earliest and loveliest flowers of spring, and *is known by the proper name of Houstonia* (*!*) but its more common and pretty name among our young flower gatherers is *Spring beauty* (*!!*) and from its nature and appearance is frequently called *Innocence* (*!!!*)." Did anybody ever print three more ludicrous blunders in one sentence? No country girl of ten years old would mistake the Innocence for the Spring Beauty. You could as easily make her believe a peony to be a rose. They are about as much alike as a tulip and a daffodil. Yet the astute writer describes *both* flowers without discovering the difference, even quoting the verse, and does not discover that though it describes the Innocence, it does not count straight nor paint right for the Spring Beauty or *Claytonia* (instead of Houstonia), which has *five* red and white petals *always*.

> "It comes when wakes the pleasant spring,
> When first the earth is green,
> *Four white and pale blue leaves it has*
> *With yellow heart between!*"

Again hear the astounding discovery made by this discriminating naturalist. "It opens its bright blossoms in the evening, and continues until late in the morning, and then closes them during the rest of the day." As neither the Houstonia nor Claytonia answers to that impeachment, doubtless the Primrose or Morning Glory was also confounded with them in the author's mind. But hear now! Here come the "ginniwine scientifikils."

"Well do I remember on a fair day in spring, taking my botany book and going to the grove to gather and analyse the wild flowers that there bespotted the quiet landscape. This (which?) is the first I plucked by my path, whose history I traced and whose name I learned."

And the *last*, I should think, by the accuracy displayed in distinguishing family differences in plants. The two or three thus amalgamated in this floral olla podrida do not belong to the same species, genus, or natural order—are totally dissimilar in roots, tops, leaves, stems, flowers, and seed; not even being of the same color or manner of growth, nor the same size, the Spring Beauty being thrice the diminutive dimensions of the other, and growing scattered about everywhere, while the Innocence grows in patches, making a regular little sod of its own. So much for sentimental botany sweetened with the "Language of flowers." If the Diadem's piety, which it professes to teach, is as bogus as its botany, we wish it a speedy death, and a funeral sermon as profound as its own scientific researches; and would suggest as a text the one found on the 47th page of the Diadem, but *not* in the Bible:—"Consider the lilies of the field, they sew not, neither do they spin!"

If these books, and the counterfeit "language" they teach, had not usurped the place of the beautiful science on which they have grown like parasites, they would not be worth the trouble of chastising. But when sensible people come to the conclusion that botany amounts to little more than the language of flowers, and that language such idiotic gibberish as I have quoted from, surely it is time to enter a protest against such usurpation and desecration.

No! Ivy does *not* say "Festivity," nor "we will not part!" as these Lexicons tell us, but it says "Old ruins —crumbling walls—and churchyard stones!" and suggests bats and owls.

To a sentimentalist of this sort, laurel will say "Glory" according to the books, while to the farmer it more emphatically says only "Dead sheep!" Yet there is hardly a bumpkin in all the woods who does not pause to admire the laurels when they are in bloom. It matters nothing to him that it is not laurel at all, but masquerading in a borrowed name. It is just as beautiful as though it were a true laurel and did not kill sheep. I once heard a drunken and stupid stage driver wax eloquent in describing the June dress of a thirty mile laurel patch he was driving through, between the Susquehanna and Allegany rivers. It was mid November, but the woods were green with the vivid verdure of the undergrowth of Kalmias, and the duskier overgrowth of pines.

"It's jest a fair *sight* in June!" he said, "jest a great posy bed all the way. There ain't no garden like it. All the roses in Pennsylvany ain't a primin' to it, I tell *you!* Jest come here in June, and you'll say you never saw flowers afore!"

But then he digressed immediately from the beautiful to the murderous, to tell how deer sometimes got caught by the horns in the laurel brush, and became an easy prey to hunters or wolves.

Doubtless the laurel, like everything else, varies its speech to suit the notions of its listener. To me the laurel speaks less of glory, or murder, or beauty, than of liberty. It is untamably wild. Try to cultivate it, and it dies. Leave it in its native haunts, or cultivate around it, even if you do not touch it, and it dies. Transplant it ever so carefully, into a place apparently every way congenial to it, and it grows no more. The only condition you can supply to make it thrive, is absolute liberty. Let it alone, and make everything else do likewise, and it thrives. Meddle with it, and it dies. This is emphatic language, and yet no sentimentalist sets down "Liberty or Death!" as the motto of the Kalmia. I see in the Horticultural books, ways laid down for the cultivation of this shrub, but I never knew one to succeed in raising them, though I have known many to try. Many of the orchidaceæ have this same deadly antipathy to civilization, and can only be enjoyed by the human race in the place where nature puts them. Do but loosen the earth about their roots, and the fibres begin to shrink and wither. Their strange fantastic flowers wilt away, and the buds blight. They say very plainly, though not loudly, "Mind your own business!" and if we disregard the admonition, they die to rid themselves of our meddlesome impertinence. Yet, as they are very charming, unobtrusive flowers, a sentimentalist, if he interpreted them at all, would make them say, "Shrinking timidity," or some other such twaddle, if he did not ignore these characteristics altogether, and give them some absurdly incongruous speech to make, as if some nine years' old urchin in short jacket should set about making Mark Antony's oration over the body of Cæsar.

There is no richer native order in the Northern States than the Ericaceæ, and none with a more classic-like mellowness of language. Abounding in graceful and varnished evergreens, in aromatic and spicy bitters, and tenderly colored and fragrant flowers, it has a fuller vocabulary and a more flexible tongue than most vegetable tribes possess. I have already interpreted the speech of the Kalmia, so far as I understand it, but how differently speaks the wonderfully fragrant flowers of the leafless Azalea. It has no winter-loving foliage, to glitter through the drifting snows, or to welcome the caresses of the north wind, but an affluence of deep rosy cups for the bouquets of spring. It comes just after the rock-loving "Arbutus" trails its garlands among the dead last year's leaves, over the sunny side of wooded hills, breathing out its odorous life in incense. Just now, in mid July, one of the smallest and prettiest of the tribe, the pipsissawa or prince's pine, is in its glory; its sharply notched, deeply veined, glossy evergreen leaves making a background which throws its clustered, rosy, waxen bells, or rather Lilliputian saucers, into exquisite relief above.

These each have a speech of their own; each dainty, exquisite, and ambrosial, but different and individual enough. The Azalea gives one an impression of fragility quite foreign to the other two. It is a banquet flower, enjoying to the full what it has to-day, and is nothing to-morrow. It says, "Let us eat and drink, for to-morrow we die!" The trailing arbutus says as plainly through the invisible sources of its fragrance, "Seek and ye shall find!" The pipsissawa crowns itself with flowers at midsummer, and, like Barnaby Rudge's raven, cries "Never say die!" to everybody who will listen. There are multitudes of carefully cultivated exotics not half so beautiful or piquant as this wild flower, that half our florists have never seen. The plant is well known to empirics as a medicine, and to garland makers, at Christmas, as a beautiful decoration, but who expects it to bloom, or who has seen its flowers? The somewhat similar, but not half so pretty, wintergreen, is a very attractive plant to children where it grows, not only the spicy scarlet berries but the entire young plant being eaten by them. One hardly expects eatables to speak to the eater in any but gustatory language, yet it is a significant fact that the tongue is both the organ of taste and of speech.

Many persons having a vague understanding of these vegetable languages, and feeling their unwritten poetry quivering like music through their souls, imagine that any scientific knowledge of plants would dispel the poetic afflatus, distilled like an etherial essence in the hidden nectaries of flowers, and the secret chambers that brew the volatile oils. They are afraid of being disenchanted by huge words, and disillusions by prosaic facts. I fancy those persons would prefer music to be sung in a foreign language, so that the effect of the sound need not be disturbed by a comprehension of the words.

There are a class of botanists of the mechanical sort, who are to science what the patent-note singing books are to music—mechanical, soulless anatomists of parts, and dictionary-like vocabularies of technology, who perceive no laws, conceive of no causes nor forces not laid down in the text-books: mere shallow smatterers and quacks, who look at a plant as they would at a new chimney, to see how it is built, and when its external structure is discovered, and its habitat indicated by a vegetable directory, and its name and lineage discovered in unintelligible Anglo-Græco-Latin, are satisfied, especially if its chemical qualities recommend it to the cook or the physician. These scientific bores disgust those who *love* flowers and trees for the sake of their beauty and poetry; and so, because some scientific prattlers are stupid and prosy, they think that the science of plants itself is unlovely and undesirable.

If our only conception of artistic music were the thrumming of a novice on some tormented and wailing piano, we should hate the name of musical art, and feel the very calling it music took some of the sweetness out of the unwritten music of nature. But if no tyro ever drummed discord on a piano, no Thalberg ever would have drawn divine harmony from its keys. We must have some prophetic notion of the end from the beginning, in order to make the beginning endurable. Many would-be musicians never get beyond the dreary regions of thumped and rattling keys; but there are a few who reach that region where Orpheus dwelt, and move the hills themselves with the utterance of their melodies. There must be beginners in everything; but should the ignorance of beginners frighten off or disgust the lovers of truth and nature from the science which is a key to both? There is a world of difference between using a dictionary to find the meaning of words, and hunting up words on purpose to find them in the dictionary; and that is the difference between the quack naturalist and the real one. The true lover of poetry, when reading some soul-stirring poem, may come to a strange word, and not find the poem lose its significance and beauty if he should refer to a lexicon for the meaning of the new expression. A pedant may read a work on purpose to mouth its most obscure phrases, after he has hunted out their meaning in a vocabulary. So a true naturalist will find many hard words in the beautiful poem of Nature, and will search out their meanings, not for the love of *words*, but for the love of *meanings;* while the quack will swallow the technical glossary for the sake of putting it forth by mouthfuls. But we need not let that spoil nature for us, nor need those whose ears are attuned to the sweet accords of the speech of growing things allow these pedantic and sentimental "Flora's Lexicons" to prevent us from learning, at first hand, the real "language of flowers."

This night might have been taken for an ideal model of one for "camping out." The soft breeze which wound up the river dispelled the mosquitoes which so tormented us the night before, and yet scarce broke the deep hush of midnight by its faint and drowsy breathing. Our mother was very near us then—nearer than we can ever approach her by daylight. Her breath breathed into our nostrils. The slow pulses from her deep fountained vital currents throbbed in our arteries, and her life became our life. Thank God! for the taintless highland air that made such a bivouac safe, even for those who had never tried it before!

But even the keen relish of a new experience could not long keep awake a weary crew, who knew that after that night's lullaby was over a new day with new experiences would call them to action again. Before day began to be visible to our eyes the wood-thrushes, who slept nearer the sky than we, began to break the silence with their gushing songs. Shade after shade of darkness ebbed away before the dawn, till at last a golden glory crowned the hill tops, and showed them not so far off as they seemed in the sombre twilight of the evening. How near the sky itself looked here! Only a pine's length above the taller tree tops. The white fragments of cloud that floated between the sky and hills seemed in danger of tearing their silvery robes on the loftiest branches, and the wide-winged hawk that sailed among them, bathed in them without going beyond our sight. The sky looks immensely farther off from the open country, and the clouds are miles away. They are both too far off for neighbors and friends, such as they are to the highlanders. Here morning did not seem to flame upon us from a distant pavilion, but arose melodiously from some breezy couch among the eastern hills, and, shaking out a thousand perfumes from her dew-dropping locks, smiled on our camp and kissed us.

We arose and trimmed ourselves, and ate our breakfasts, and chattered and sang, like the other happy creatures about us. Morning is the time to laugh and be glad, as evening is the time to muse and be serenely happy; so, as we had ended one day naturally, we began the next naturally also, and laughed and were glad.

The promised boat came, early, but we were ready and waiting, and embarked for the regions below with alacrity and glowing expectations. We neither rowed nor sailed, but were drifted and *poled* along, seated upon perches like so many turkeys gone to roost, on as many flour barrels cushioned with our own bedding. The boat was a scow—a sort of shallow, square contrivance for carrying loads, and this was loaded heavily. The draught was very inconsiderable, but the channel of the stream on the ripples was more trifling still, and we accordingly stuck fast several times. When stuck only at one end, we all went to the other extremity of the boat, and thus lightening that part, floated off again; but when hung by the middle, that kind of tactics would not do. Overboard went a lot of bipeds into the warm flashing water, and seizing the four corners of the craft with vigorous hands, and applying strong shoulders to them, swung the great thing, thus lightened and urged, around broadside to the current, and free again. In again leaped passengers and crew with a joyful burst of congratulations, and a considerable extent of dripping garments to be again hung over the sides of those barrels to dry in the wind and sun. Perhaps the soaked individuals within them tempted a siege of rheumatism or pneumonia by their many wettings, but no such distress came in consequence.

We had deep water, and slow, dream-like floating on a mill pond up to the dam, which was high and had the "slash" on besides, and was consequently impassable. A delay of several hours occurred here, as the boat had to be unloaded, and taken around the island, which formed one shore of the pond, and floated and dragged through the narrow and shallow channel on the other side, and reloaded at the foot of the dam. Unloading our own trumpery and piling it up for reloading, we proceeded down the river on foot. There was a house here in the process of erection, and loitering behind my companions, in a deep hollow, to examine some plants, I became the unintentional repository of a confab among the carpenters, as to whether we were Chinese, gipsies, or other outlandish folk; but having secured the dalibardas I was in search of, I emerged from my hole and lost the conclusion of the matter, but went on, mentally resolving that as we would not be Chinamen on any terms we must be gipsies, and might as well go to fortune-telling at once.

The river banks are high and steep, and often nearly perpendicular, with a gravelly beach at the foot at low water. Along this gravelly river's margin are sprinkled many cold clear springs of water, close to the river's edge. A slight freshet covers them all with turbid water from the hills and surrounding country, but now that the river was low and warm, these beautiful little wells we scooped out with our hands to make drinking places, were an overflowing blessing to the thirsty. The banks were covered with uncivilized verdure very delightful to rummage amongst. Ripening whortleberries made a blue glimmer among the sparse and patchy foliage of the low undershrub which bore them, both making a harmonious picture on the russet hemlock "muck" in which they grew. The chestnuts were in full bloom, and no pity for nutless squirrels next fall deterred us from adding some of the longest and most graceful of their plumy flowers to our collections. Diervillas out of their common yellow uniform, sported lively orange red blossoms on the steep declivities. Two dainty species of Apocynum spread their delicate foliage and pink blossoms over the hot sand, and among the duskier greenery above. The ragged border of hemlocks fringing the top of the bank bounded our horizon when we were down by the water, but narrow as our range of vision was there was plenty to see. The spring flood-tide of bloom was over, as well as the fullest flush of the summer flowers, but enough yet blossomed to reward research and continually whet our appetites for more. The seed growth of the deep-woods plants, too, was a continual feast of discovery to most of us. Having spent in woods like these more summers than I care to count just now, this forest growth was well nigh as familiar as the foliage in a kitchen garden; and a two years' residence in prairie land had made me hunger and thirst for these woods and waters so keenly that I went "maundering" about, greeting my old friends beside every log and under every bank. They made fun of me, and dubbed me the "Dictionary" that day for the unconscious pedantry of mouthing so many jaw-breaking names for innocent little weeds, which were wholly unknown to the Greeks whose language has been appropriated so much in naming them. These hideous names, which serve as so many scarecrows to frighten off lazy schoolboys and girls from this most beautiful study the green Earth affords, have become as pleasant as household words to me, from constant use. In order to learn them, I contracted a habit of talking to plants in my solitary walks, and calling them by these names, and, despite the absurdity of the thing, I continually betrayed this habit among my fellow travellers; so I was the "Dictionary" Ah, well! one could be much worse a book, if not a less poetical one.

Elvira took a second lesson in shooting to-day, and so far improved upon her practice at the mosquito hunt of the day before, that she brought down a woodpecker from the top of a hemlock, much to her own astonishment.

When the boat landed us on the right bank of the stream a few miles below, it was at one of those riverside fountains which gushed out in the shade of a green though fallen tree, which was bathing its head, with its "terms inverted," in the Allegany. Near this the Indians had made a footing in the precipitous bank, up which we climbed, and after a half hour's reconnoitring, selected a spot overlooking the river, and went to building our camp. We encamped under a beech tree, using a stout horizontal limb, which stretched towards the river, for the ridge pole of our wigwam, supporting its outer end with a stake. Against this ridge pole we leaned boards and slabs, which were plentiful in this vicinity, and here we had a picturesque camp, opening riverward and commanding a prospect so lovely in its rural simplicity and quiet that our artist took possession of it immediately. We were on the outside of a curve, just above Jemmison's Bend, and could see a long panorama of inverted hills and woods done in crystal far up and down the "Beautiful River"—for the Senecas continue the name Ohio, or rather Ohe-yu, or "Beautiful River," up to the very head of the Allegany, which is a name they never applied to the river.

I have heard, though I do not know on what authority, that Allegany signifies "Head of the mighty." If it does, it is in allusion to the mountains of that name, from which so many rivers rise.

The river near our camp afforded us a delightful opportunity for swimming and playing in the water—a pastime never out of date—an enjoyment new every summer. The Seneca children seemed to be amphibious, as though they were akin to porpoises. One little fellow played riding on horseback, by sitting astride a floating slab and propelling it with his feet.

A descendant of Mary Jemmison, the "white woman," lived a quarter of a mile off, across the river, and his family crossed the river in a log canoe, which lay at the landing, several times a day, to get cold drinking water at the spring under our bank. A woman with two children came oftenest; and it was a marvellous sight to see her send that narrow craft as straight as an arrow across the rapid stream. She used a setting pole, and stood up in the middle of the canoe, and would give it such an impetus before it reached the deep channel where her pole was too short to touch bottom, that it would dart across the swift current without perceptibly swerving in its course. The gentlemen of our party tried to learn the trick, but bungled and failed every time, to the great amusement of the skilful boatwoman, whose clear musical laugh it was worth being laughed at to hear.

The next day was Saturday, and as we had a written invitation to a grand Sabbath-School picnic a few miles down the river, our company gladly availed themselves of the opportunity of seeing the people together.

The old rulership of the chiefs is superseded by a republican form of government, and the picturesqueness of manner and attire are passing away. They are obliged to be civilized in self-defence. If the white people about them were not the most barbarous of their race, the Indians would probably prize civilization more. As it is, the exemplifications of enlightenment they are in the habit of associating with, are not such as present very powerful attractions to an unvitiated taste. At a world's fair of scalliwags, I think some of the white scamps on the river in the Reservation would take the premium for being the most worthless, drunken, foulmouthed blackguards possible to find. Added to this chronic nuisance which empties itself into the river from Ellicottville and the other towns near the Reservation, especially on Sundays, to fish and play the ruffian or ribald clown generally, there is a periodical nuisance of a more business-like and less idle character, but which is morally no better for the Senecas. All the up-river lumber region uses the river for a market road, and whenever there is a freshet of sufficient magnitude to make the stream navigable for rafts, which happens twice or thrice a year, the region below is deluged with lumbermen, who at home are not generally the best possible specimens of Christian civilization, but who leave their greatest stock of good behavior at home when they go down the river. They are oftener than otherwise rough, profane, and drunken. Now the railroad, with its degraded hangers-on, crosses the Reservation, and runs near it for many miles. There are, however, some permanent residents not to be reckoned in this category. On asking a Seneca if our baggage would be safe alone at our camp, he said, "I am afraid not. The Indians and the neighbors (the white residents just mentioned) won't touch any thing, but there are some Irishmen and white fellows over here that hadn't better find them alone." However, we left them alone more than once, and lost nothing, so it is supposable that only the Indians and the "neighbors" found them exposed.

Mr. Purse, who invited us to the festival, besides being President of the Seneca Nation at that time, is a Baptist preacher, a Sunday-School teacher, plays in the brass band, teaches singing school, translates hymns and sets them to music, and sometimes composes both—supervises all the printing done in the language—is the prosecutor of those who sell whiskey to the Senecas, makes speeches to the whites in good English—and is now engaged with some three or four others in translating the New Testament into their tongue. In his wild younger days he travelled over the greater part of North America with a band of musicians, and now seems to be making the best use of all he learned in his peregrinations. Appreciating the advantages of true civilization, as well as the difficulty of bringing them home to his people where they are, and as they are, by any means hitherto used, he has now on foot a new project partly executed, for the more thorough instruction of a part, at least, of the Senecas. He has selected thirty young people of both sexes, whom he desires to get into Christian families in decent neighborhoods, to learn the arts and ways of civilized life, and to get a good school education besides. He has already got places for several, and hopes, by sending out his people to learn, to do what a few teachers among them never could accomplish. He hopes they may be able to maintain themselves by their labor, and thus be no expense either to the whites or themselves. Those already out are doing well.

The white "neighbors" mingled in the festivities, and helped get up the refreshments, and the whole affair went off swimmingly; but with the exception of dusky faces and the strange language of the dusky speakers, it was like any Sunday-School celebration, animated by a brass band and a feast. True, the gala dresses were somewhat in aboriginal taste, though not in pattern, but it was as civilized as anybody's celebration.

Mr. Purse, whose Indian name I wish I could remember, was, of course, one of the principal orators of the day, and spoke in his grave, earnest, energetic way, that well nigh made his Seneca tongue intelligible to Yankee ears. The closing honors were awarded to our captain, whose speech seemed to please the Indians as much as the Indians had pleased us, one delighted old man asserting in very bad English, that this was "The best speak he ever did heard!" How much of it he understood is problematical. Owing to their appreciation of that speech, however, our friend was invited to preach to them the next day, and had the offer of a canoe to take him to meeting. Of course he went. I rather think he would have done it if he had never done such a thing before, for the poetic inspiration of the place and people seemed to endow him with the spirit to do his best without any trouble.

Before morning, however, there were two considerable dampers cast upon our enthusiasm. The first was the discovery of a complicated sectarian feud existing among them; the nation being divided into the "Old Style" and the "Christian" parties, and the latter being subdivided into the Presbyterian and Baptist parties.

The second damper was a shower in the night, which put our frail wigwam to a test it could not bear. Even the Baptist portion of our party had to put up with being sprinkled that night. It did not rain in the morning, so, as we did not wish to appear to take sides in their feuds, half of us went to the Presbyterian meeting close by the camp, and the other half went down to the picnic ground among the Baptists.

After church we had a taste of the good manners of some stray Ellicottvillians, who were fishing in the river. Some of them were in boats and some on shore, and two of the latter came tearing through our premises, and swung themselves down the bank by the ridge pole of our house, pulling the mansion down after them, just as it began to rain again in good earnest.

We didn't like it, but we couldn't help it, and so we ran about looking for shelter, like a nest of ants whose home has been unroofed by taking away the flat stone they had encamped under. Back in the woods a piece, just in the edge of a raspberry patch, stood a log cabin covered with boards, unfinished, and never used. There was a hole cut for a door, but no door in it, and a hole in the roof for smoke to go out, but no chimney. The spaces between the logs were unchinked, and there was no flooring. A stray cow seemed to have lodged there occasionally, but we removed all evidence of her former presence, and strewed the floor thickly with hemlock boughs, and having removed ourselves into it, bag and baggage, dried ourselves before a roaring fire built under that hole in the roof. It rained all that night and the next day, and made quite a little freshet, dirtying the clear Allegany sadly, and washing away our springs. The floodwood came down at such a rate that we could not cross the river, and a projected visit to the aged Patriarch Blacksnake, the nephew of the renowned Red Jacket, had to be foregone. That was an irreparable omission, for within the year he died, at the age of one hundred and twenty-three years. The oldest of the nation now living, can remember him as a middle-aged man when they were little children.

As I said before, the gourmands eschewed our company, but for all that, as we were neither spiders, nor snakes, nor angels, we had to eat occasionally, and as our cupboard grew as bare as Mother Hubbard's, we began to think about converting raw materials into eatables—for even savages *cook*, and the utmost simplicity of living we could attain to would not make Allegany mountain trout eatable without fire. I believe I am aboriginal enough, and hate dish-washing badly enough to make me satisfied to roast my fish on a stick and eat it from a leaf, and then throw away my gridiron and plate to save washing. Thus extremes should meet, for I would at once be as simple as an anchorite, and as luxurious as the Emperor of Japan, whose dishes are broken as soon as used, so that nobody else shall profane them by eating after him, nor he suffer the indignity of eating twice from the same dish. But my companions thought that neither the august prestige of the potentate's example, nor the thriftless laziness of mine, could keep the fish thus roasted from tasting smoky, nor add the desired flavor of salt and butter, so there must be so much of the abjured arts of civilization used as sufficed to fry fish. Mrs. A. and Susie had provided all sorts of provender, even down to pots of butter and bags of salt, but nobody had brought a frying-pan, nor even a "dish kettle," which one of our ingenious countrywomen has averred is convertible to every culinary purpose whatever, from a coffee-pot to an oven. Some of our Seneca neighbors had sent us some peas, and here were peas to boil and fish to fry, and a pint tin basin and a self-sealing fruit can, holding a quart, were our utmost resources. With a stove there would have been little difficulty, but it rained pitilessly, and our fire was a huge heap of burning wood on the ground, under that corresponding huge hole in the roof, which let out all the smoke which went straight up, and let in all the rain which fell straight down upon it. The wind, too, took an occasional whim of introducing variety in the camp, by varying the position of the column of smoke and fire from its perpendicular to all sorts of angles between that and the horizontal, and sometimes whirling it into a spiral shape, which however pretty to look at, took up as inconvenient and uncivil an amount of room, and was something less approachable than a crinoline in a waltz.

Our captain had taken advantage of the morning rain, when trout like to be caught and always bite, and had come home with a fine string of them, and was nothing loth to stand drying his bedraggled length before the log heap; but as he stood looking forlorn enough, a bright idea was working under the touzeled exterior of his caput. He went out and cut a slender beech sapling, some inch and a half through, and stood up gravely again, whittling industriously at his stick. Off fell the leaf-laden branches, hissing and snapping into the fire, as he lopped them away, and finally the sapling seemed a club about eight feet long, split at one end. Into this split, the edge of the pint basin was insinuated, and held by the elastic wood, like Milo's fingers in the log trap he caught himself in, and lo! there was a Lilliputian frying-pan with a Brobdignag handle, and the biggest Professor from our Alma Mater transformed himself into the *chef cuisinier* of our expedition. In went some butter and a coiled up fish, and our extempore skillet was held over the fire by a muscular arm, and watched by a metaphysical eye, as though it was the manliest occupation in the world!

Our captain is a good surveyor, and would make a good engineer they say—is a good mathematician and a good logician—he teaches well and he preaches well—is six feet high—has big whiskers, and a bass voice—and added to all this he improvises frying-pans, and fries fish capitally. And he looks just as dignified and philosophic doing that as at anything I ever saw him do. He forbad our calling him "Professor A ——" fifty times, but we forgot it every time, and he seemed just as professional in one place and plight as another. He might stand ever so long *en dishabille* with the water dripping out of his red flannel shirt, and upon his head from the leaky roof, and fry trout one at a time for our breakfast, and we would call him "Professor" in good faith while he was at it. It was as good as forty lectures on the dignity of labor. I wondered then, more than ever, where people ever get the absurd notion of talking about "refined" and "vulgar," or "masculine and "feminine" employments. It sounds as ridiculous as the French way of calling knives masculine and forks feminine. My knives are no more masculine than my forks. Elvira's shooting was as feminine as her curls, and the Professor's cooking as manly as his beard. Susie was as lady-like when she washed up our scanty dinner service in the frying-pan, as when she played and sang in her mother's parlor, and our captain never was more a gentleman in the pulpit or on the rostrum than when making that frying-pan. I believe he could darn his own stockings, sew on his own buttons, or perhaps possibly he might even sell tape behind a counter, or sit a la Turc on a shopboard and stitch jackets, and not bate one jot of his manhood. I exulted as I thought how those brawny hands had carried that massive head up from a poor and obscure place by their power to work, and thought that so long as the "mud-sills" of the Bridge of State are made of such right royal, and at the same time such stubbornly democratic timber as this, the Bridge *cannot* sink—no, not even under the weight of a "latter-day" President, Cabinet, and two whole Houses of Congress! But here I've got clear into both Houses of Congress and our peas ain't cooked yet! There's velocity for you! I think I might help Mr. Field and his corps about that troublesome Atlantic Telegraph! But as my efforts in that line can hardly be appreciated yet, and as I can prove a capacity for cooking peas, I will leave that cable in its sea-salt pickle, and its owners in their monetary pickle, and tell you how I dulled our poet's jack-knife, scraping the cement off the old tomato can, in order to make a dinner pot of it. Somebody says that kind of cement is poison—the which I don't know—but it is uncomely and unsavory, so it had to be scraped and melted and scoured off, and then when the peas were put in, and the lid on, we had neither bail to hang it by, nor handle to hold it by, nor stove to set it on. But we did not mind that, for Necessity, the mother of Invention, had several of her children with her in our camp. The Professor could make frying-pans and the Poet could manufacture shovels. A stray clap-board, gleaned from a pile of drift-wood by the river, was duly hewn and whittled into shape and did duty as a shovel, and was, from a respectful distance, thrust into our little volcano, and robbing it of a pile of coals, our extempore dinner pot was set thereon, and cooked away with profound satisfaction.

The Europeans laugh at the American propensity for whittling, and no doubt it is ridiculous sometimes, but it is a natural consequence of plenty of wood and few tools, and our experience proved jack-knives to be a great institution.

We would ask you to dinner if you would not be scandalized at the scantiness of our dinner service. One table-knife and fork, three jack-knives and a pen-knife, two spoons and two plates, and, it must be confessed, our *dinner-pot* and *frying-pan* added to our two tin drinking cups, was the whole. It would have been better for each to have carried a private spoon in a side-pocket, but we didn't think of it, and so we waited for each other when we had "spoon victuals," and at other times did what our ancestors did before forks were invented—used five-pronged implements like those Mother Eve managed with native grace in the bowers of Paradise.

Nobody can tell till he has tried it, how much fun there is in doing without things. It is really astonishing how many things people can do without, when they can't help it. Altogether the merriest meals I ever helped to eat, were thus merry not on account of what we *had*, but on account of what we had *not*. Some absurd and irremediable omission provokes mirth, where people are not silly enough to feel chagrined at the poverty or the blunder. If we had made our outfit perfect, and had forgotten nothing, we should not have had so many things to laugh at, and should probably have enjoyed it less.

Next day the heavens "stopped crying," and we went up the river in a wagon, and slept in a house on a floor that night. The air seemed very close after the airy chambers we had been occupying.

There is a "Rock City" on a ridge of hill near Great Valley, where conglomerate rocks are piled many feet high for several miles in length. Here in a crevice of the rocks part of our company encamped the ensuing night. Feathery ferns and mosses tufted the rocks and trees, but the shade was too deep for any luxuriance of undergrowth. Some of us got a bit of a ducking in crossing Great Valley Creek to get there. The bridge was gone, and the water so deep that it floated our wagon-box and wetted our bedding. It got dried, however, and none of us got drowned.

Here my western enemy, the ague, which had tormented me through Illinois "and back again," like the champion in the old game, who couldn't succeed in dodging the old witch on the way to Bileybright, came on with double force and drove me home. When I lay sick and fevered by the Mississippi, drinking warm wiggling water, I felt as though the cool springs and clear air of these hills would cure me at once—but it took more than one week.

On my way to the dépöt I visited one more curiosity, and one yet unexplained. This was the "breathing well." It looks at the top like any ordinary well dug for domestic purposes in a farm-house door-yard, and this was its original design, but after digging some thirty feet they came to a strong current of air, but no water. They dug no further, but put a sort of penstock in the top of the well, with a large tin whistle fitted into it, which answers every purpose of a barometer. When the air is light the current rushes out of the well with great force, blowing away anything which is laid over the hole, and blowing the whistle like a young steam engine. If the air without is heavy, the well draws in its breath with the same energy.

The cause of this phenomenon is unknown, being theoretically supposed to be owing to some hypothetical cavern down there somewhere. It is worth the investigation of the curious.

My companions pursued their way to visit the new coal-mines a few miles south of here in Pennsylvania, but as a true historian's tale must stop where his knowledge does, I must pause here. If you are as sorry as I was, I am glad.

THE NATURE CURE.—FOR THE BODY.

I ONCE heard an enthusiastic Geologist say, that when he was a confirmed and almost hopeless invalid, he took daily doses of the Tertiary Formation for one season, and found himself well. He had been a close student, and had contracted an inveterate dyspepsia, which, he was told, exercise would cure. His first attempt was a signal failure. Every day he started for a walk, in the spirit of swallowing a dose of pills, and walked along absorbed in his books, or else thinking about his ailments, saying to himself continually, "This is to cure the dyspepsia; this rod is to leave the dyspepsia one more rod behind; so much exercise vs. so much dyspepsia," &c. &c. But he found after a six weeks' trial, that giving his distresses an airing only increased them, and the more he tired himself, trying to walk off his indigestion, the more it would not be walked off.

A Naturalist spent a few days with him, and his thoughts became keenly interested in the earth-history recorded beneath his feet. With the whole energy of his excitable temperament he threw himself into the study of Geology, and in pursuing the thread of its revelations among the hills and along watercourses, down wells, and up precipices, his aches and ails vanished.

The keen out-of-door interest he had gained, carried out his thoughts from the sphere of their chronic habits, and with every breath of air that filled his lungs, the now bounding pulses poured the once creeping blood into their vitalizing laboratory.

The pulsations of the aroused soul stirred the half stagnant currents of the body, and renewed their life in the free air. His disease was lost off and forgotten. He no longer thought of inspecting his tongue at the looking glass, nor feeling his pulse, nor keeping an account of the variations of his appetite. He had something far more interesting to think of, and the added relish it gave to the minutiæ of life was enjoyed without analysis or reflection. He did not ask of himself how much such or such an organ had been benefited, in order to enable it to afford him such and such a pleasure more than usual. He had always led some such forced and methodical existence before, but now he led a spontaneous life, and the long dormant seeds of many joys began to swell and sprout within him. He was no longer the pasteboard globe revolving on pivots within its wooden horizon and brazen meridian, but a planet rolling through space, with infinite blue on every hand, and the kindly influence of many moons and stars drawing him in an eccentric and wavering orbit around his central sun.

How could he remember to have the dyspepsia then?

Half the diseases of mankind would get well of themselves if they were not so petted and thought about. Many a common cold has been worried into consumption and the grave, by constant attention and apprehension. We are perpetually warned against being careless or negligent of our physical disorders, and the warning is sometimes needed; but it as often happens that the continual reminding us of it aggravates the difficulty, as it happens that care alleviates it.

We generally find what we look for in this world, and in this field as truly as anywhere. If we look into every coughing fit for the seeds of bronchitis or consumption, we will find them after a while, and if we watch them incessantly to see them germinate, by and by sure enough, they will begin to grow, when, if we had not used our anxiety and attention as a sort of hot-house and forcing-bed for them, they never would have vegetated. I have known a spinal disease to be induced by constantly thinking about and expecting it, and daily looking over one's shoulders into a glass to detect a curvature. People go mad, sometimes, for fear they will be crazy.

They say that Pan, the genius of the woods, is dead, and that his mistress, Nature, went out of fashion long ago, and is quite obsolete in polite circles now; yet if there is a *pathy* in the world, worthy of trust, it is that taught in their school.

Hydropathy, as practised by Naiades and Undines, must be delightful in warm weather, but the difficulty of finding the physicians at home when you call, makes its efficacy uncertain. But Pan, who may have died, and come to life again, for aught I know, is alive now, and always at home. Kill him ever so many times, he won't stay dead. He has more lives than a cat, and what is better, he has the power of transferring his vital tenacity to his patients. He is the greatest of physicians, although mythology says nothing about it, and though his bill is not the longest. He has other modes of advertising the superiority of their mode of cure, than making magnificent charges. There is no quackery about the Nature Cure. It never orders nauseous poisons to be forced into the stomach, in the expectation that they will find some road to the liver or pancreas, or other ungetatable places, when they are out of order. To be sure, other doctors do it, but that does not hinder the venomous medicine from finding the wrong road just as easily as the right one, and blundering into the wrong gland, or some other place, and go to curing what is not sick, just as fast as what is.

This is a different system. The man who took the tertiary strata did not swallow them, though he took them in ether—a sort of ether that keeps best without bottling up or corking. Yet it was an internal application, acting with great force on the brain and all the vital organs, although so agreeable to take, that it soon became more delightful than any feast, the only disagreeable results being the fatigue consequent upon so much exercise. But what sweet sleep followed those days of toil, and what pleasant dreams grew out of the invigorating atmosphere he breathed all day long!

An invalid who does not feel quite equal to the task of carrying a stone hammer and a basket of rocks over the hills, as a sanitary measure, can of course take a different prescription. For instance, I have known a professional man, addicted to liver complaint and other bad habits, take half an acre of cabbages with excellent effect.

Another, who ought to have sued his ancestors for damages for imposing on him a poor, rickety constitution, contrived to renovate and make it quite passably useful, by the application of five-and-twenty varieties of rosebushes, with pinks and geraniums *ad libitum.*

That this mode of cure is always accessible, at least to all country people, is one of its chief recommendations. Skilled physicians of any other school are so often ungetatable, that the children of poverty must submit to be experimented upon by tyros and quacks, or go unattended through sickness, unless they have faith to try the out-of-door treatment, which, in a great majority of cases, if taken soon enough, will perform a cure; and if it fails, and the patient dies, he has at least the satisfaction of dying a natural death, without the assistance of the Faculty.

There is much truth in the old proverb, "An ounce of prevention is worth a pound of cure." All physicians of every school will say, that any disease is much easier cured, if taken in time, than if deferred as long as possible. Some diseases of the most formidable character—contagious, hideous, and often fatal—may be wholly averted, if proper steps be taken before exposure to infection; and greatly modified after it, so as to be neither frightful nor dangerous, if properly treated from the moment of their inception—as for instance the small-pox. Let it alone, and it will kill or disfigure its victim. Vaccinate or diet, and he will come out whole and sound in most cases. We claim no miraculous power of restoration is to be found, even in the purest air of the mountain-tops; but a vigorous preventive, and a salutary, healing influence for those inclined to disease, or already partially ailing, is plentiful and free. Mother Nature herself would not take a far-gone consumptive into her hospital. When a man gets too ill to get out of doors, it is too late for him to try that method of cure, and often too late for any method to do him any good. If you had a friend going to be hanged, you would not wait until the rope was about his neck, before you started to get him reprieved; but when our friends sicken, we think it is time enough to do something when they can no longer sit up or eat their meals. The wonder is, not that people so often die young, but that anybody lives to be old. It exhibits an almost incredible tenacity of life, for a sedentary person to get well after a fit of illness.

Divers spasmodic efforts to restore the regime of Nature in the medical world, have risen and waned, alternately astonishing the world by their audacity, and amusing it by their absurdities. But the passion for humbug and mock erudition which has always more or less characterized this profession, and the fact that it is and has been a *profession* instead of a study, a science, and an art, have made these new schools ruin themselves by their Brobdignag pretensions and Lilliputian achievements. The little fire of truth under this witches' cauldron of medicaments and theories, has kept the uncanny broth bubbling, and these ephemeral attempts at reform have been but the bursting bubbles on its surface.

A European school of Empirics once professed to have discovered that the breath of cattle was the most curative atmosphere an invalid could inhale, and prescribed in certain chronic diseases that the patient should be laid on a pallet in front of the feeding racks of the cattle, in order to breathe in, and otherwise absorb, the hygienic odors of the kine.

People have been sent to the tops of mountains to regain the health they might have kept if they had only visited the mountain tops on their own feet sooner, instead of waiting to use the feet of mules, when their own were no longer available.

They have been sent to the bottom of caverns for the nitrous air, that was to heal the lungs that were decaying because they had never had enough of any sort of air. Pale students and professional men who have made their offices the ante-rooms to the Court of Death, by breathing dust and carbonic acid, and the fumes of ink and tobacco, from one year's end to the other, run under ground in Mammoth cave to escape yet a little longer the final putting under ground, which they have not dreaded enough to avoid sooner.

Merchants, from whose abused nostrils the everlasting smells of the shop and store seldom have a chance to subside, take costly pilgrimages to far off springs, to drink abominable waters, to undo the evil that never would have befallen them if they had breathed more clean air, and drunk more pure water, as they went along.

Ladies who have squeezed the breath of life out of their lungs, and admitted the chills of death to their bare arms and shoulders, and might-as-well-be bare feet, go into the sea brine with the lobsters, after the health that grows thriftily in the kitchen garden and on every hillside.

Thrifty housewives who fry their vital juices out over the cooking stove, and fill their lungs with evaporated dinners, and their stomachs with indigestible delicacies, generally take to patent medicines and kindred abominations to purge them of their unnecessary ailments. They, seldomer than any other class, think they can afford time, or have strength to take any part in the Nature cure the others try to get at.

I sometimes think women like to be sick, they take such capital care to destroy any health they may have inherited. They think frail health is interesting. Don't the poets write delightfully maudlin verses about beautiful consumptives coughing away their forlorn existence in the most sentimental manner possible? Romantic young ladies, with rosy cheeks and bright eyes, feel complimented at having some simpleton insinuate that theirs is the radiant eye and hectic flush of incipient phthisis. Health is vulgar. No fashionable lady will own to so prosaic a condition. Did you ever hear one say right heartily, without any modification of *ifs* or *buts*, that she was in excellent health? Not she! If nothing under the sun ails her, she will sooner improvise a pain in her side, or her head, or her little finger, or her great toe, rather than be supposed to have no ailment whatever. She thinks it prettier to be sick; and for fear she might be well if she let herself alone, she exposes her extremities to the vicissitudes of the atmosphere, and crushes in her vitals with jean and whalebones, and loads the lower viscera and vertebrae with a young camel's load of dry goods and hardware to drag down her life and impede locomotion. Of course no animal thus loaded, hoppled, cramped, and exposed, can apply the nature cure without altering her habits and defying fashion, and this few women are brave enough to do. She prides herself on being a coward ("timid" is the euphemistic phrase), for that is almost as necessary as ill health to make a lady interesting. "She might as well be out of the world, as out of fashion," and so she puts herself out of the world as expeditiously as the prevailing mode of suicide prescribes. No matter for the Divine command, "Thou shalt not kill!" Fashion says, "Kill yourselves and your children by slow torture!" and Goddess Fashion is obeyed, and the Lord's commandment broken. *Murder* is a stern word with which to confront and accuse these delicate dames, but what less have they done? If they like not my speech, I commend them to the third chapter of the poems of the Prophet Isaiah, and pass on.

Young ladies in their teens, who have resolution enough to defy parental advice and authority, and run away with some perfumed popinjay, to be "married in haste to repent at leisure," have not moral force nor physical courage enough to put on a pair of calfskin shoes, and a dress short and loose enough to climb hills in without stepping on the skirts, or gasping for the breath they have no room for in their dress waists; and then, thus equipped, to take rambles, and even scrambles, in places rough and smooth in pursuit of those objects of beauty or curiosity, of scientific or artistic interest, which are always accessible in the woods and fields.

No, indeed! *Scientific young ladies are never the heroines of novels!* They are absurd and unfashionable, and are supposed always to have ragged blue stockings, inky fingers, and dowdy hair. They are believed to be predestined to be old maids, or the worthless wives of victimized husbands, and the slatternly mothers of neglected children. All the oracles of romance have shown that women never take to science or art, except from starvation or disappointment.

I am not going to dispute her prey with Romance, for her votaries are too destitute of gumption to be brought to their senses by anything short of the sharpest realities of life, and then it is often too late for them to undo the absurd mischiefs they have done to themselves.

Romance makes the confectionery of life, but whoever undertakes to live on sweetmeats must pay the penalty of such folly.

Half the American people are women, or will be, if they live long enough, but how is the nation to stand alone with so flimsy a "better half" as these women make? The mass of American women work as hard or harder than the men, but they accomplish far less, because they work at an immense disadvantage. Even the women of the country do themselves up in such a way that if they were perfect Sampsons they could not do much but tire themselves out in trying to do. Put Sampson himself in hoops and long skirts, and set him to carrying up full pans of milk from the cellar dairy room, and see how he would make out. Set him to hoeing the kitchen garden, or weeding onion beds, and after he had tripped himself up a few times by treading on his "flowing, graceful drapery," and after thereby ripping it loose from the waist, and tearing and muddying it in sundry places, see if he wouldn't swear, or "think swear" at least. If he had had on a long-shouldered dress, and shoulder braces, when he met the lion, I'm thinking the bees might have found another hive than the lion's body, unless the lion had worn corsets too, in which case the man might have killed him.

Suppose you send him out in a lady's rig to do an errand, in a high wind, and see how, with all sails set, he could make headway against the storm. Let him feel all sorts of things flapping, scourging, entangling, twisting, jerking, snapping, and dragging away at every part of him, and he would cry out, like Cain, that his punishment was greater than he could bear. Call woman the "weaker vessel" indeed! If she were not stronger than Sampson, and more patient than Job, she *could* not, and *would* not submit to the fatigues and inconveniences of such habits, and if she were not sillier than a goose she would not be the slave of raiment that she is.

But what with fashions, and fashion-plate magazine literature, and customs and prejudice, it will be hard work to induce young women, who are the very class who need it most, to take that out-door exercise that their health and highest intelligence demands.

The assertion that the highest intelligence needs exercise of the body in fresh air, in order for its development, may be disputed. Those whose idea of intellectuality are all connected with pallor and leanness will cite all manners of cripples and invalids, to show how energetically the mind may work in a disordered body. I know that Scott and Byron were lame, but I know, too, that they walked and swam as few sound men do. I know that Elizabeth Browning is feeble, and that she wrote better when she was well, and that her sea-side rambles have left their impress on her poetry. Charlotte Bronte inherited a consumption, and was "stunted" by starvation in childhood, but if she had not habitually taken longer walks over the bleak Yorkshire moors than an American lady would think possible, we should never have had Jane Eyre, nor heard of its author. Harriet Martineau, though incurably diseased and deaf, has written the best guide-book for the wild and rugged lake scenery amid which she lives that has ever been written. Because, not only was she more gifted than other guide-book authors, but she was practically more familiar with what she describes.

These are the sick ones, but the well ones are in the majority. Most sickly geniuses die before they arrive at such celebrity, or indeed before they get out of their teens. Think of Bayard Taylor or Ida Pffeifer hobbling round the world on crutches, or scaring the sea nymphs of Pacific or Indian seas, with the ghostly pallor of their dyspeptic visages, or the hectic gleam of sunken eyes! Imagine Banvard, or Church, or Rosa Bonheur painting the Father of Waters, the Heart of the Andes, or the Horse Fair from luxurious carriages, in embroidered slippers and kid gloves! If Hiram Powers had remained an indoor man in a Cincinnati warehouse, or Harriet Hosmer a parlor ornament or kitchen utensil in her Massachusetts home, do you think that we should have ever had the Greek Slave or Zenobia?

It took a sea-side cave for a study, and an audience of waves, to train a Demosthenes of old, and a hunter's life among Virginia's hills to make a Patrick Henry of modern times. The life of the wandering "flower-girl of the Boulevards" ripened into Rachel, the Queen of Tragedy; and how much had the wild air of the Polar region to do in developing the gifts of that Swedish song-bird, Jenny Lind?

It takes a well oxygenized blood to stimulate the brain to healthy action. Carbonic acid mixed with the purple currents of life, does not agree with the nucleus of the nervous system. Genius itself must succumb to half-vitalized blood and a vitiated atmosphere. The "midnight lamp" addles more brains than it enlightens. There is philosophy in the old notion that moonlight crazes people, that is, if they stay awake of nights to look at it. Nature forbids over activity as well as sloth, and haunting about at uncanny hours, as well as staying shut up at all hours.

These two absurd notions, that it is not pretty to be robust and rosy, and that it is pretty to try to superinduce beauty or genius by making owls of ourselves in turning night into day, or by rat-like inhabiting unventilated and unwholesome quarters, seem, in stating, too ridiculous and foolish to require a rebuff or refutation; and yet there are multitudes of our fellow-creatures and *creaturesses* acting upon them daily.

Why else do our ruddy, stalwart farmer boys seek to be doctors and clerks, as though goods and drugs smelled sweeter than fresh mould or clover sod? They think white hands measuring tape and assorting fancy buttons, or fingering deranged pulses, are much more stylish and "genteel," than honest sunbrowned hands. They think a varnished and useless cane a more interesting and elegant occupant of their digits than rake or plough handles would be. And why else do fat and rosy young ladies drink vinegar and eat chalk and tea-grounds to diminish their flesh and color, if they do not think health and strength undesirable? Or why do would-be geniuses keep late hours, and roll their eyes ominously, if they do not receive these notions?

We may ignore this large and increasing class of dunces, and treat only of those we generally think of as the more sensible class, who are the vigorous doers of the land, and yet we shall not find them obeying the laws of life any better than the others. They may plead the excuse that they are wearing out for some purpose, and not from sheer folly; but I know not that the Divine command, "Thou shalt not kill!" makes any exceptions in favor of those who work themselves to death.

Mankind seem not to have learned that however much the body may bear of abuse, it is the worse for every ill thing done to it. A man may be sick and get well again, yet that illness has shortened his life. He expends his health recklessly, and says that he is strong, and can bear the strain of his unnatural living; but though the machinery may endure "by reason of strength even to fourscore years," the fresh vigor of life that might have been retained, has gone long before that, and a man of seventy now, is older than Adam was at six hundred.

Even wise men say that the times demand a rapidity of action which cannot be sustained except by unnatural stimulus and expenditure of vital energy. The spirit of the times demands that we should hurry ourselves to death, eating, drinking, sleeping, and working, with a deadly activity, which prevents any of these things being done thoroughly.

Why this scrambling haste? Have you not time to live *well* while you are about it? We Americans are in such a desperate hurry we have not time even to chew our dinners before they are swallowed. We can't wait for the cars to stop before we jump upon the platform, nor wait for the steamboat to be moored before we leap across a nice place in which to drown, in order to get on board or on shore. We count it wasted time to do slowly what it is possible to do rapidly even to live. The deliberate philosophy has few votaries among our countrymen. Because man is not a ruminating animal, he seems to think he needs no time for chewing anything but gum or tobacco, and these he chews as he runs.

To move leisurely along the ways of existence, in order to let the experiences of life have full time to mature their proper fruit in the character, is considered too slow for the Age; and so this double high pressure engine Age drives everybody as if life were a general racecourse, where he who gets through quickest wins the stakes. Few of us like even slow, solemn, and grand music; a waltz or quickstep better pleases us.

All our inventions are for hurrying up the processes known before, or contriving more rapid ones to supersede them; and we partake of the character of our inventions. We cannot wait even for hides to be tanned into leather by the tedious processes of the old time, but rot them by more rapid and less substantial and efficacious methods. It is the same with most of our fabrics. Dispatch, rather than durability and excellence, is the prevailing idea in the manufacture of everything from college Baccalaureates to shoe pegs.

We have thus created a false and flimsy taste among us, which, from having few substantial things which would endure the wear and tear of time, has made us come to like the incessant change; and if we do happen to have anything last beyond the ephemeral existence of its contemporaries, we get tired of it, and demolish it on purpose. People get tired of durable garments,—have had them so long they are sick of them. They get out of conceit of beautiful houses which have outlasted the style in vogue when they were built, and pull down the excellent old house only because it is out of fashion, and build another not as good but more fashionable, in its place.

Even old men and women are out of fashion, and if there is now and then one, they are the odd exceptions and not the rule. The unavoidable evil, when it does come, is concealed as far as possible by all manner of arts and affectations, as though old age were no longer honorable.

Thus we have destroyed our faculty of veneration by destroying what is venerable among us. As we measure age, our nation itself will soon be old and in its dotage, and must make way for some newer style of government, or some old style come round new again, as our styles of dress are renewed.

This absurd and destructive racing system must either rush us into this and other undesirable results and catastrophes, or be abated and cooled off, and taught to go slower. So long as we are proud of our rapidity there is little hope of mending it. We must meet some overwhelming disaster or disgrace, to humiliate us, if we will learn wisdom from nothing but experience. If nothing but rushing into the gap of an open drawbridge will convince our fast engineers that the car of state has no wings, why we must have a general crash, as on our railroads.

But those of us who love smooth waters and quiet scenes, can at least set before our fast countrymen and women the dangers of haste, and the open and inviting fields of health and recreation, where the overtasked and broken down may regain in fresh and quiet pursuits, that strength they have destroyed by their over hasty living.

It is pitiful to see the pale and hollow cheeks, and the languid and drooping figures, of so many of our young men and women whose vital energies were originally sufficient to have carried them bravely on to a green old age, if they had not been used up by false living, and so left them old and worn out in their prime. There is no artificial process of renovation. The decayed teeth may be replaced, but the torture of the process will leave its pain brand on the face, like a wrinkle of age. The grey hair may be dyed, but the white roots will be continually showing at the parting. The body's bones may be hidden by drapery and wadding, but the sallow skin hanging loose upon the skeleton of the hands and neck, betrays the waste of life within. The skin may be daubed with whiting and paint, only to be a mockery without, and a slowly absorbed and deadly poison within. A full beard may hide a sunken cheek, but not a sunken eye. Cloves and cardamom seeds, and orris root or perfumery, may conceal a pestilent breath, but not the cough that belongs to it. Poetry may decorate disease, but can never make it pleasant or beautiful.

If one can be well, it is morally wrong to be sick. We deserve better of ourselves than to murder our good bodies that the dear Lord gave us to take care of our souls with. If we do it carelessly and ignorantly, still the penalty falls upon us. There is a death-penalty attached to these physical transgressions, and whoso breaks the laws of life is a self-murderer.

There are things worth dying for, but how many of us who transgress the life laws do it in a cause worth the cost?

You whose flying feet crush the life out of so many trooping moments as they come, and whirl yourselves dizzily through stifling rooms into the cold air, that cuts its way through your heated lungs, to the very seat of life—you who have giddily danced the "Dance of Death," was your pleasure worth the life ye forfeited for it?

And you, dismal Ascetics, who believe that God only loves long faces, and who evolve ponderous sermons out of your bile and dyspepsia, darkening the beautiful sunshine and the joyous smiles of life with your raven shadow, and drowning its gushing laughter and song with your croaking and groans,—when the acrid acidity of your lives corrodes your health away, and brings you down to death,—do you think that your gall and vinegar temper, which you mistook for piety, was worth the life it cost?

Servants of Fashion! Has your capricious mistress ever repaid you for the corns she has pinched on your toes, the breath she has crushed from your lungs, the curves she has given your spines, the perpetual inconvenience and expense she has cost you, and above all for the wasted hours and shortened life she has caused?

Slaves of Luxury and Appetite! Where are the rewards of your fearful sacrifices? The ancients cast their infants into the fiery embrace of their brazen Moloch as an offering, but you have consumed yourselves by slow tortures as living, yet dying, sacrifices to your terrible Idols! Item by item have your bodies been diseased and tortured to pay the penalty of your devotion to Appetite. Gout and dyspepsia, tooth-ache and liver-complaint, cholera and delirium tremens, apoplexy and heart disease, nervous torment and lunacy, are the rewards of your indulgence in the pleasures of Appetite, Luxury, and Indolence; and the loss of life, the extreme penalty the body can pay, is the end of it in this world. You have eaten and drunken and revelled, at the price of your lives. Were your debauches worth their cost?

Devotees of wealth, what reward have ye? Lives of toil and bondage, to gather, *what?* You have denied yourselves the free gifts of air and exercise perhaps, or have worked beyond your strength, and have accumulated *dust* for those who outlive you to quarrel about! You have deferred life until rich enough to live, and now you die. Have you not sold life too cheaply?

Ambition's votaries, who have wrought with all your force to make a niche in the temple of Fame, where you might dwell and be admired of mankind, what reward have ye? Disappointment and envy and death! Are you paid for your lives?

Oh! all ye whose lives have been a feverish strife after some imaginary good, come and slake your thirst and cool your fever in the clear waters of the many springs among the hills. Let nature's breath breathe into your failing lungs the healing of her own. Let the serenity of Nature steal upon your unrest, and give you her tranquillity. Allow the vigor of her unfailing forces to renew your lives. Let the great pulses of her ever beating heart throb with your own.

Thus even hereditary taints of disease in the constitution may be overcome in a measure, and the burden of pain earned by those who come before us, and not by ourselves, may be removed.

I should find it hard to pardon my ancestors for my existence in the flesh, if what they had given me were leprous, or scrofulous, or consumptive, or predisposed to insanity or special vice. Yet I see many of my fellow creatures have existence thrust upon them thus burdened and cursed. We see distorted, deformed, diseased, and crippled wretches, who are so from no choice of their own, but because they were the infant sacrifices devoted to the Moloch of their parents' appetites and passions, or their cruelties and mistakes.

These unfortunates seldom have the mental or moral vigor sufficient to reverse the process of destruction begun by their progenitors, and by living in accordance with the natural laws, rescue themselves from a progressive march from torment to torment in the abused flesh. Too commonly they go on as their fathers did, and increase instead of arresting the inherited plague. The son of the drunkard, who has inherited the hideous disease of his father, feeds this morbid appetite with the poison which adds to its craving, and he is sooner a drunkard, and more incurable, than his parent. The child of gross and vicious parents inherits not only the physical consequences of their transgression, but a passional bias in the same direction with the parents, and will be liable to re-enact their deeds, and so suffer an aggravated and double penalty. But those whose mental perceptions and moral force are sufficient to turn them to the simple and efficacious regimen prescribed by Natural Law, will be likely to find healing and health in it.

For the remaining few invalids, not included in the foregoing, who are ill by accident, and not in consequence of the culpable ignorance or wickedness of any body, there is as much hope in the Nature cure as in any. Not that Nature will set broken bones or dislocated joints, but she will have very much to do with curing when they are set.

I wish not to undervalue any profession, but to give nature her due, and she is above all physicians. Moreover, the claims of the suffering are paramount to the claims of all the professions under the sun. Doctors, as such, have no claims on the public, except as they heal the sick better than they would be healed without them.

The plea that we should patronize the various hotels, shops, schools, artizans, members of the professions, etc., in order to *sustain them*, seems to me to be arrant twaddle. What claim have they to sustenance? They are for certain uses, and in so far as they are of use, use them, but no further. If they are not sufficiently useful to command support by their benefits, let them fall. Any institution, or in general any individual who cannot stand alone, with at most a little propping to begin with, might, better fall than pretend to stand. Who props up the seedling pine in the forest to make it grow straight, and not fall down from weakness? If one cannot grow from the centre outward, self-balanced, and fast by his own roots, and be helped to grow only for the indispensable fruit he bears, he had better relinquish his ambition to be a tree, and be only what his own power and capacities would make him. Let your supernumerary doctors and lawyers, etc., take to ditching or ploughing, or some of the trades that are never full, and in which it is hardly possible to have a superfluity. If there were no other way to prevent doctors from starving, except to use their services when we didn't want them, there might be some charity in patronizing them, especially if that were cheaper than paying a pauper tax to support them in another, but analogous way. But as circumstances are, where is the need? Let us appeal to the Great Physician out of doors, and let the M.D.'s do as they like, if they do not interfere with us. Surely they have no right to feel aggrieved if we only let them alone. It is often the very best thing we can do by anybody, to let them be. We will give them full opportunity to mind their own business, by minding ours. If we need them we will ask for them, and such patronage will not be a charity nor a humiliation to the man who feels he does only what is needed, and is paid only for what he has done. Good physicians agree that they are only nature's aids; and bad ones, who set up in opposition to her, are those we wish to see driven into some less destructive employment.

Come then into the fields and watch the bright-eyed toads catch flies, and learn of them how to trill the French R; or go down by the marshes and look for "Pete," whom you hear so many little frogs calling for. Perhaps he is the very doctor who can cure you. Or go into the woods and look for bright spotted lizards under the stones and old chunks of rotten wood, or watch the woodpecker listening like the veriest eavesdropper, at all the cracks of the dead tree, to hear what the grubs are saying, ere he opens the door with his bill, and arrests the incipient beetle at his dinner. Go into the beech woods and watch the squirrel stuff his cheeks with the triangular nuts, and go frisking off to his hole, without seeming in the least encumbered by his loaded pouches. See the pigeons scratch up the dead leaves like so many of the most beautiful and graceful of hens, laying bare the sprouting beechnuts they wish to carry to their young; or if you are able, follow them to their nesting place, and see acres and miles of these beautiful creatures at home among their fat squabby babies, and try to estimate the amount of life stirring the air there, and you will begin to feel as if you were absorbing the superabundant and overflowing vitality into your own systems.

"Go to grass." It will do you good; if it does not, it is an error easily repented of, and a course easily abandoned, and one that will leave no poison in the system nor sting in the conscience. But try it, and rest assured, on the experience of those who have tried the experiment, that greens will be good for you, if properly taken.

I hope yet to see invalids giving their full faith only to that school of medicine whose basis of operations is as broad as all-out-of-doors.

THE NATURE CURE—FOR THE MIND.

PEOPLE, especially the young, often get a diseased appetite for excitement and artificial amusements, and what is called society, to the exclusion of far better things; not because the taste was originally depraved, but because life has presented to them no real and abiding interest.

It is common to hear such persons declare in hours of reaction and discontent, that they find nothing worth living for; they find no pursuit worth all the toil and trouble it costs, and no aim worthy of life. The pleasures they devour last but a little while, and continual change and new excitements are craved with perpetual hunger, which no amount of feeding ever satisfies—the only satiety being spiritual dyspepsia and disgust.

Men usually seek refuge in business from the hungry needs of their inner selves, and try to quench their soul's perennial thirst with eager draughts of gain. Like Midas, they seek to turn all that they touch into gold; and when it is done, they find themselves like him, starving at their own banquets, because gold is neither food nor drink.

Women commonly seek their refuge in matrimony, only to find themselves called upon to feed instead of being fed; and behold their garners are empty and their fountains dry—they perish of thirst and starvation of mind and heart, and others perish with them; so we see the world peopled with the famished skeletons of men and women, who seem, Prometheus-like, to be for ever renewed, but to be for ever devoured by the craving vultures of immortal want and undying need.

To convince us that this condition is abnormal and unnatural, we need only look at nature's ever fresh tranquillity.

To cure it in its chronic state, is only possible to Infinite Grace—a physician not often applied to for that purpose.

Wisdom will avert the disease in the beginning, or find it easily cured in its earlier stages, the chief difficulty being to get other people to allow the patient to take the prescription, not because it is difficult or dangerous, but because the few who admit the existence of the disease despise the simplicity of the remedy. If it were some great thing, they would believe in it, but what is the use of leaving the big Pharpar and Abana to go and wash in that little Jordan?

But the mass deny the existence of the disease, and the necessity of cure, just as madmen rave that they are not mad.

Others say that this is a want for which there is no supply, a disease for which there is no cure. Man was born to hunger and thirst unfed—to suffer unrelieved. This life must be barren and starved, that in the life to come we may have more abundance. It is a doom from which there is no appeal—a fate from which there is no escape.

They might as well say that children were made on purpose to cry, and that their mothers ought to starve them in the cradle, to make them "laugh and grow fat" when they are men.

It seems to be a sort of general belief that people ought to be treated by the rule of contrary, and supplied with exactly the things they do not want. When a *body* is naked they give it sermons, and when it is hungry, spiritual advice; but when the *soul* is starving scholars feed it with Latin and Metaphysics, common folks offer it pork and potatoes, or perhaps a new coat or bonnet, and too often religionists only give it creeds and formulas and beliefs, instead of the bread and water of life; and when the *mind* is hungry they give it no meat, when athirst they give it no drink, when sick and in prison they minister not unto it, because they cannot comprehend the perishing need.

To know, is as positive and keen a want of the mind as hunger is of the body, or grace of the guilty soul.

As everybody recognises the necessity of food for the body, and in this country most people believe in the need of aliment for the moral part of the human trinity, I pass to that part of our hungering which the great mass of the people think may go unfed.

They can see no use in so much learning. What will be done with it? If one can read, and write, and cypher, he has enough to do business, and be "respectable," and what more would he have? There are not wanting those in our midst who think it a waste of time and money to buy and read books; and some of our next-door neighbors think that knowledge tends to knavery and pride, and eschew it accordingly. Said one of this class to a poor school-teacher, "Why in the world do you keep pinching yourself, and working and saving, just to send yourself to school? You know enough to teach school now, you ought to be satisfied and *be like other folks.*" "But what shall I *do*, if I do not go to school?" asked the girl. "Oh! dress yourself up and *catch a man*, and get married and settle down, and have a home of your own." "And what then?" was the query, not asked often enough, by the way, but the answer was only that she was to do then as others did. A daughter of one of this sort of people importuned for the privilege of mental improvement, but was stopped with the information that "she knew enough to keep house," and that "the *boys wouldn't like her any better if she knew ever so much.*" "She *could not make better johnny-cakes* if she should go to school or study ever so much"—which must have been immensely consoling to a hungry mind.

One of our neighbors, a rich farmer, had a son who worked faithfully for his father until he was twenty years old, with only a scanty chance of going occasionally to the district school. At last the youth, with a keen sense of his own deficiencies in learning, besought his parent to allow him to go *one term* to a sort of an Academy some twenty miles distant. Consent was grudgingly wrung by hard teazing, from the father, and a very spare wardrobe and *a single shilling* furnished the young man with all the outfit and means his generous parent thought worth while to "throw away" on "book learning." And this was not in Tartary nor "Borio-boola Gha," nor in the dark ages, but in the United States, in the middle of the nineteenth century. This kind of mental starvation is terribly common too, though these people would feel grievously slandered if accused of starving their families; but what better have they done? Unfortunately, too many of these hungry minds get used to being starved after a while; and though they do not quite die, they pine and wither away, until what might have been vigorous and majestic intellects are stunted dwarfs, nourished with meagerest scraps of food, and watered with vain regrets.

Set before the mass of people, any form of knowledge not absolutely necessary to use in the getting of food and shelter, and they ask, "Of what use is it?" This question must be answered according to their own ideas of use, if they are answered at all. It may be as well worth while to answer the ignorant according to his ignorance, as to "answer a fool according to his folly," as scripture recommends. You can put on a wise look, and give them what will be an answer to them. Tell them that geology will teach men how to find coal and lime, and metals; that botany will teach what plants are poisonous, and what are eatable, and what are good for medicine; that entomology will teach where to look for cutworms, how to get rid of caterpillars, and how to manage bees; astronomy will tell when the moon changes and when the tide will be high, and will help men make almanacs. Mathematics teach how to reckon interest, and, by the aid of chemistry, gunpowder, and shaving soap, and other such things have been invented or discovered.

What if this truth be not *the* truth to you? No matter! It is of no use to tell your truth, that knowledge, like virtue, is worth having for its own sake, rather than for the sake of its uses. They will not understand nor believe you. And yet, like the man born blind, who will not believe in nor understand light, they are suffering for want of that in which they disbelieve.

They believe all these needs of which I have spoken, to be only imaginary wants, and as for the imagination itself, it is supposed to be too vague and unreal a thing to have any real needs of its own.

Just what Jesus says we are to take no thought for—"what we shall eat, and what we shall drink, and wherewithal we shall be clothed," are the only things we *do* take thought for and reckon valuable, except money, that treasure we are warned *not* to lay up in this rusty, moth-eaten, and thievish world.

"As a man thinketh, so is he." Our thoughts are us. What we *think*, we *are*. What we *know*, we *are*. What we learn becomes part of our minds. That which we remember who shall teach us to forget? *Is* death the everlasting sleep unbelievers teach? Do we, Christians, *really believe* ourselves immortal? And if we *do*, where are the treasures we are accumulating to take with us into the safety of an incorruptible world?

We can take only ourselves along, without the bodies of dust, and all the wealth we can take must be part of ourselves.

Where is that wealth? We can carry the memory of houses and lands, and food and raiment, with us, but will these remembered riches make us rich in eternity? Are they any element of true happiness here?

Answer, O Midas! starving at thy own golden feast!

Will the memory of the giddy excitement of *mere* amusements make a part of that wealth? How shall thy dancing skill content thee, where feet dance no more? Or how shall the ribald jest of the circus-clown, or the airy walk of the acrobat, reach, even in this world, to thy inner discontent, or appease its cry? And canst thou carry with thee, for a single hour, a serener tranquillity, for having seen centaurs at the hippodrome, or the dizzy gyrations of the puppets at the theatre? When the lights are out—when the dance is done—the play played out—what is there left but the dregs of petty jealousies, and envyings, and heart-burnings, with all the flavor and aroma of pleasure exhaled and gone?

The most candid of pleasure-seekers only claim for such things, the power of diverting for the time the painful or wearisome current of every day thoughts and feelings; or the office of filling up with its manufactured mirth and dazzle, the oft-recurring hiatus in life, which else would be empty and bare.

Amusements have no higher office than diversion; they never can be a true *interest* and *object* in anything worthy to be called life. And yet how many intellects have asked for bread, and had these stones given them! Whatever of value these things may have in their own place, they never can be satisfactorily substituted for the substance of worthier things.

But find a real interest in those strata of truths which lie below all the mud and soil of what is temporary and artificial, below the crust of conventionalism, or the quicksands of the relative and speculative, where men get lost as in a quagmire; ground your interest on what is positive and accessible, and life itself becomes a pleasure.

He who sups on sweets will find his mouth bitter in the morning, but a simpler food will nourish without any after penalty; so if you gorge the mind, or its servant, the memory, with unsuitable diet, they will get a distaste for food from indigestion.

This state is worse than that of famine. It is not the stuffing process of the ordinary educational mills, called schools, which is to supply the wants of the hungry minds, supersede the vulgar and paltry amusements of the day, or overthrow the dynasty of the Blondins and Blitzes.

The ordinary educational programme is too much like a recipe in a cookery book, to be the thing needed. "A cup full of butter, a cup full of sugar, a cup full of flour, a cup full of cream, four eggs, a teaspoon full of soda, with salt and spice to your taste," is not a very essentially different prescription from, "A book full of geography, a book full of grammar, a book full of arithmetic, a book full of words to spell, an empty book to be filled with written words, a little composition, a little declamation, with singing to your taste."

To be satisfied with only this, would be as impossible as for the four years' old lad to love to go to school where his whole experience consisted in "sitting on a bench and saying A."

No traveller, however expeditious, ever expects to see all there is to be seen in the world. It is too much for him. Life is too short, and the world too long, for such an exploit. Neither does any person expect to learn all that there is to be learned in the realm of knowledge. A man who would try to farm a whole continent would have poor crops, but he who ploughs only so much as he can attend to, will have abundance. One thing thoroughly done is worth more than many slighted jobs. One field of inquiry thoroughly explored, is better than a smattering knowledge of many.

There are two Grand Divisions in the world of Learning—the Temporary and the Immortal.

Of the Temporary are all manual Arts, all of the *this world* Literature which contains no essential and undying truth, all Languages which will be useless where the thoughts will use no syntax nor lexicon in their utterance. All these have a single attribute of immortality, for the discipline they give the mind, it retains, when the learning that made it strong and astute, becomes rubbish in that future life we are preparing for, where there will be no use for such knowledges.

The Immortal Learning is that which we can not only carry with us, but *use* in eternity. These are Positive or Absolute knowledges, those truths eternally true, demonstrable by plain proof, or immediately perceivable, things which can be reached without theory or guess work—the very bones and sinews of learning, the kernel of knowledge—the for ever satisfactory. In this realm, too, lies the border land between this world's knowings and the next, where the Philosophies dwell, like the angel of the Apocalypse, with one foot on the earth, and the other on the unseen sea.

Learning is not wisdom, nor is it always knowledge. It is not what is put into the mind, but what the mind is made capable of doing and producing, which is the desideratum.

Knowledge is mental food. The mind gets hungry for the aliment it digests and appropriates to grow strong upon, just as the body feels the want of physical nutriment.

Undigested food, either physical or mental, is useless and often detrimental.

Yet knowledge is worth haying for its own sake—is worthy of love and effort for the sake of its own value. The pursuit of it is a keen pleasure, and its attainment a solid satisfaction.

Nothing is ever earnestly pursued for any sake but its own. If we, for a time, seek one thing for the sake of another, the pursuit is dropped when the object is gained, or when a shorter method of reaching it offers itself. A student may study for a diploma, and when he gets it, he forgets his wearisome learning, and takes to something he likes better. The man who stops his drams to win a wife, will take to his cups again when the foolish woman is caught; he did not love virtue for its own sake, but adopted it temporarily for another sake. The student who studies for the sake of the uses of knowledge may make a *good* scholar, but never a *great* one. All greatness must grow from an all-absorbing love for the object. The pursuit of the object becomes a passion, and the passion becomes exalted into devotion. Other considerations may be of importance, may be respected—attended to in a measure—loved second best—but the profoundest love is never forgotten, never neglected, never put off for minor things—and all other earthly things *are* minor in comparison with this.

Such devotion makes the great painter, the great musician, architect, or naturalist. The mere ordinary aimless aims of life are poor grovelling to them. The majority of people live because they were born, and have not died yet. They only live because they are alive, and do not know what else to do. Any vegetable does as much. But these others have ennobled life by their brave living. Most of us leave it ignoble enough, in our aimless twaddling which we accept as life. But try to bribe or threaten men or women who have some such object in existence—who live on purpose, and are glad they are alive—who would have chosen to be born if they had been consulted, and who live, even here, after they die. See what answer you will get of such!

The musician Haydn (I think it was) when threatened with what would be a terrible punishment to an ordinary artist—the condemnation of the popular school of music, which, they told him, forbade certain liberties which he had taken with the established rules. "But they are agreeable to the ear, and *I allow them!*" was the reply of this man, who loved his art better than he loved success. His great love conquered the arbitrary rule, and achieved success in the teeth of criticism.

"Do not paint, or you will starve!" said her friends to young Rosa, "you shall be a milliner and get rich!" But the enthusiast loved art better than money or bread, and studied beasts in the dirty shambles, rather than laces and flowers in the shop, and through her severe and toilsome obedience to the one absorbing passion of her life, she conquered obstacles, and poverty, and prejudice, and stands crowned with wealth and honors among those who would have made her a milliner.

The Emperor of France offered the most honorable place and the highest salary offered to any Professor in his empire, to the Swiss naturalist, who replied that he had not time to attend to it. A *little* man would have had no time for anything else! "You shall have any price you ask, if you will but give us a lecture!" said a lyceum committee to the same great naturalist, but, "*I have no time to waste in making money!*" was the simply sublime reply of Agassiz.

Ah! such souls live in a clearer atmosphere than the dull fogs most of us inhabit. They reckon profit and loss by a different currency, and their riches cannot be computed in gold and silver.

They have loved their love *for its own sake.*

An ordinary conventional life is so hollow, that it will sound like an empty cask, when it comes in contact with what is substantial; but let this life grow as if it were alive, and be filled with substance, and it will no longer ring with its own emptiness.

There is no languor nor ennui to enervate the mind which has taken an earnest hold of the teachings of nature.

He who is intimate with nature at home cannot be ignorant, though he may never have learned the alphabet. He has the key to a realm of treasures of which nobody will dispute the possession, for, like language, it belongs to everybody who has a mind to use it.

It is a partly explored region, with new wonders waiting to be discovered by new adventurers, and exhaustless marvels for every explorer.

Happiness, with a vital zest and interest in it, abides out of doors with nature. You can find it there plenty and cheap. The birds and waters ask no admission tickets to their free concerts, and the miracles of growth and being, surpassing all the feats of all the magicians that ever lived, are there enacted every day.

Meet a man grievously vexed with the blue devils, and tell him there is an hour's serene happiness to be got out of a certain old crumbling mossy stump, surrounded by a sorrel sod and two dead mullein stalks, and he will think you are a fool, perhaps, and go on despising your childishness, and magnifying his own despair, because John Smith's note is worth only fifty cents on a dollar.

Never mind—you are a philosopher, and can afford to be thought a fool. Why not? If it is any comfort to any care-harassed mortal to set you down among the dunces, because you are not also tormented, let him have all the poor mite of satisfaction he can get out of a conviction of your stupidity. You live at headquarters, where happiness grows spontaneously with the weeds, and you can spare such bits of pleasure to anybody.

These poor mammon-ridden wretches would, perhaps, none of them be willing to have this persecuting Old Man of the Sea, of their cares, taken off their galled shoulders, to be exchanged for the thriftless joys of the naturalist.

As soon as you begin to enter into the arcana of nature, you feel the shackles of outward customs grow loose, and the liveries of many servitudes drop off, as a bird moults its feathers. Nothing is done suddenly, for Nature has plenty of time—all the time there is—and is never in a hurry. So these relics of old bondages wear out, and fall away piecemeal, and you hardly know when you lose them, but some day perhaps others tell you they are gone. They are finding fault as they tell you, and you know they tell the truth, because the complaint does not annoy you as it would have done once. Other men's opinions of you are now of less import to you than your own. You respect yourself—you approve yourself, therefore you can afford to be censured for what is best and bravest in you—you can afford to be misunderstood. What matter if even, your friends accuse you of growing "eccentric" on purpose to gain the notice of other people's eyes! You can afford even that, for you know that very unlike causes produce like results, and forgetting that other people have eyes or tongues, sometimes produces similar effects to remembering those same eyes all the time. Human vanity will permit but few persons to believe that their opinions are matters of absolute indifference to one of their fellow mortals. They can understand how a person might make himself ridiculous for the sake of being noticed, but that that notice should be entirely forgotten and unthought of is inconceivable!

Let them flatter themselves that they have demolished you by their ridicule. If it pleases them there is that much pleasure gained to the world, and as you have not felt it, you have lost nothing.

You have changed your standard of measurement. You no longer ask; "Do they believe it?" but, "Is it right in itself?" This return to the normal standard—to the absolute truth instead of the accepted version of things, will work great transformations in any character, definite enough to deserve the name of character.

Unnoticed as the process of liberation may be, the effect of the liberty is noticeable enough. A prisoner, long in shackles, when first released, still walks as if restrained; but the artificial gait soon wears out, and liberty restores the grace of unfettered motion.

In the woods and fields, people's eyes are off, and you who had been a prisoner to their criticisms, find your limbs loose and you walk freely. By and by you get the habit of being free, even among your fellows. They were your jailors once, but their term of office has expired for ever.

Of course those people to whom liberty is offensive, will find fault. They will sneer at, and perhaps slander you. What if they do? It is their bent—let them have it! (Especially as you cannot hinder them.)

If they were in China they would be shocked at a woman's foot if it took the liberty of growing as God intended. If they lived in Malacca they would be disgusted with any man who would dare to let his teeth remain "white like a dog's." If they were among our aborigines they would be scandalized at the bad taste of any military gentleman who would prefer his own unstained complexion to the fashionable bars of red and black paint, with which their code of decorum insists they should beautify their countenances.

Of course Fashion will be against you, but what of it? Counting numbers, you will be in a very small minority; but in very truth, where is the majority greater than your Liberty, Nature, and Nature's God?

Mental liberty is a prerequisite to all deep and abiding happiness, and Nature sets you free, if you are willing to be emancipated.

When you have achieved sufficient independence to be able to make your own observations, from thence do your own reasoning, and deduce your own conclusions with tolerable accuracy, you are just prepared to commence those explorations that none have achieved before you, and that the scientific world is waiting for. Why should not you, as well as another, unlock those mysteries, and enrich mankind with new revelations of science?

There are momentous problems, as well as less significant ones, waiting for solution. A few examples of these unsolved problems will convince the dullest believer in plodding stupidity, that there are at least a few scientific facts worth discovering.

Over a large tract of our Western country, there prevails at a certain season of the year, a strange and often fatal disease among cattle, known as the milk sickness. It is evidently produced by poison, and the poison-producing tracts are so definitely marked that they can be fenced in, and the cattle kept out, and the disease ceases. But though this disease is so wide spread and so well known when it appears, and its locality so plainly discoverable, no one has yet been able, with any reasonable show of demonstration, to tell the cause of this remarkable fatality. The inclosed noxious localities have been examined, inch by inch, for they are generally quite limited in extent, and yet nothing has been discovered there, not common to all the rest of the region round about.

Now the stupidest Hoosier that ever thought "larnin'" a humbug, would be glad to know the cause of this plague, especially if, as is often the case, the knowledge of the cause would indicate the kind of remedy needed.

In this case superficial and unphilosophical investigations have proved to be worthless. The only index to the probable cause, yet found, is the declarations of the only men of science who have looked into it. These are physicians, who declare the disease to have every appearance of being produced by a mineral poison. The farmers, who had before convinced themselves that if it was produced by any vegetable agent, it was an invisible one, ploughed up and examined the plague spots, without result. There were no chemists among them to test the soil, nor explorers to investigate the wherefore of those mineral exhalations, if such they were, making their appearance at one season more than another.

This problem, of pecuniary and sanitary as well as scientific interest, remains to be solved.

When will some naturalist give us the *modus operandi* of the fungus called smut, in grain, and discover its remedy?

Who will demonstrate the cause of the potato disease, and devise a preventive?

A reasonable acquaintance with natural science would prevent some disastrous mistakes. Men have beggared themselves in search of coal and other minerals, in places where any tolerable geologist could have foretold certain failure, since he would know that such minerals are *never* found in such places.

I cite these few examples from multitudes I might mention, to show that mere utilitarians, of the coarsest type, have a bread and butter interest in the growth of scientific knowledge, and that the ploddingest plodders will gain a dollar and cent advantage from the researches of those whose thriftless studies they could see no use in.

Thinkers will see that the far better and truer reasons for the study of Nature, I have given first.

"This is all very well," says one, "for those who have time to attend to it, but working people have no time for such studies."

Let us look at this excuse for a moment.

We do not expect a woman who does her own housework, and sewing for a husband and seven children, beside workmen and boarders, to be very much at leisure for intellectual pursuits, nor would we expect to find a devotee of science in a man with a sick wife and innumerable responsibilities, little and great, to provide for. The mere fact that such people contrive to live *at all*, proves how much can be got out of twenty-four hours per day.

See how many useless and foolish things get done, and reckon if the useful and healthful pursuit of knowledge would not occupy the same time far more pleasantly and profitably.

When we, the working people, make such excuses for ourselves, we justify the insolent effrontery of those aristocrats who claim that it is necessary to have a privileged class to do the learning and thinking, leaving another class to perform all labor; for labor, they say, is incompatible with study and thought.

Indignant denials of our inferiority are only ridiculous, when *we* declare ourselves to be unable to make ourselves the peers of any class we please.

Have you forgotten Hugh Miller, the Scotch quarryman's apprentice, who studied the revelations Nature made to him in his vocation, and thus became the world-renowned geologist? He tells not only the result, but the process of his learning, and shows how his love of knowledge was a perennial enjoyment and blessing to him—a safeguard against the seductions of vice and extravagance—an incentive to industry and virtue—and, though he does not say it, we know it was a rich gift to the world.

Indeed his whole career goes to show that physical labor is a help to mental growth and health, for when the growing importance of his intellectual labors and discoveries caused him to devote himself exclusively to study, abandoning the manual labor of his earlier years, the overtaxed mind gave way, and the man went mad and died.

The nice balance of the physical and mental forces was lost, and all was lost. His greatness grew out of his using his mind and his hands in connexion.

Any farmer's boy or girl has as much opportunity as he, and as much time. He observed as he worked. So can any of us who live in the country. We can study Nature as we work in the garden and the field. What few books we need, cost less than we expend for the so-called pleasures of which I have spoken.

We afford those things we care most for. If we care most for dress and show, we will find ways to afford aliment for our vanity. If we care most for money we will contrive to accumulate it. If we care most for the opinions of others, we will devote time and energy to winning their favor. If we love moral and spiritual things best, we will not stint our devotions and our souls' culture to a few hours on Sunday. If we care more for our minds than our bodies, our minds will be ministered to. If we choose, we may be learned and even wise.

Nature is not an exacting mistress. She will yield her treasures to one who gives her only the odd scraps of half worn hours, and the fag end of holidays. She is patient, and for ever new. To the weary and heart-worn, or to the empty soul, she is an untiring solace—a faithful friend. Her lessons are food and rest, and like a fountain in a desert place, or "the shadow of a great rock in a weary land."

THE PERFECTION OF THE NATURAL.

IN our often arbitrary use of the terms "good" and a "bad," "perfect" and "imperfect," we betray a faith, instinctive and unconquerable, in the existence of a positive standard of perfection, to or from which all our comparisons are made.

If beauty or goodness were but comparative qualities, our language would not need, and never would have had a superlative form of expression. But the simple fact that mankind, even in the earlier ages, felt the necessity of some mode of expressing the different degrees of excellence, from a lower up to a higher, and *the highest*, shows the existence of a common instinct of the race recognising the ideal of the perfect, and the aspirations of mankind have ever been directed towards this ideal.

Slowly, as through the lapse of ages Truth fed the hungry multitude by the hands of her chosen philosophers and prophets, has the world recognised a nearer approach to the absolute standard.

The soul of Nature stretches out her hands in her great yearning towards the better and the higher, and though in her eagerness she grasps too often a spurious good—a delusive shadow of better things—still, though cheated and mistaken, and crowned with a thorny and stinging experience, her hunger and thirst after perfection proves the existence of a supply suited to her wants.

Nowhere in God's universe is there any natural want given, without a suitable supply, prepared somewhere in nature, to meet it. The lips and speech—the ear and sound—the eye and light—the heaving lungs and the inbreathed air, in their adaptation to each other are but a few from myriads of examples and proofs of this truth.

Nature is young yet, with childhood's instinctive unrest and foreknowledge of its own predestined growth and maturity, and though her hopes are sometimes wild and extravagant, the slow coming perfection of her life shall as far exceed her present ideal, as the realities of life transcend the visionary castles in the air built for its future by the child.

Begin where we will, in the lower, material world, we find this same tendency from a lower towards a higher; we find that there exists an ideal type of perfection for every grade of being.

From the slippery mud that clogs our wagon wheels, and scarce serves to make the commonest brick, through the finer material of which the chemist forms his crucible, and that which the potter moulds into shapes to suit his will, to that highest type of argillaceous existence—the porcelain clay, which labor and fire transform into creations of such beauty, that human pride loves to add them to its stores of brittle magnificence.

From the yellow-grey desert sand, to the crystal and stores of gems,—from dingy charcoal dust, to diamond sparks and the "mountain of light" to which Brahmins paid divine honors.

From the lime we plaster on our Avails, up through the rich marbles and alabaster—that build fountains and tombs—palaces and monuments, and growing life-like under the sculptor's hands, give down from age to age the artist's ideas of physical perfection. Through the strange machinery of animal skeletons, and the varying beauty of many-tinted shells to the coral trees growing with the instinct of animal life, working and breathing, as it builds island walls from old ocean's floor up to the sunlight.

From all of these advances towards perfection among existences below consciousness or volition, we infer inevitably the progress of the very elements themselves, towards the most perfect development possible to their natures; and the continued transition from the mineral to the vegetable and animal kingdoms, by becoming aliment to supply some of their needs in the mysterious process of nutrition and growth, evinces a tendency in lower to grow up and pass into higher states of existence.

The miracle of Adam's creation from the dust of the earth is renewed in the growth of every living creature, since the earth and the elements are converted in the laboratory of vegetable existences, into the food which is again transformed, in the alembic of the body, into materials for forming the body itself. Like Adam, every man is made of dust.

This great sermon on the transfiguration of matter, preached by the stones during all the six thousand years of man's existence, has been heard in different ages, by different disciples of nature, who, while accepting the outward truth so far as they understood it, gave always some mythical and hidden significance to the plain speech of their teacher, evolving by the magic of poetry, from stubborn facts, beautiful fictitious theories by which to solve eternity's problem of the origin, the existence and the desiiny of spirit.

Mrs. Child brings up the old idea of the growth of spirit out of matter, or rather its growth *through* matter, in her "Poet's dream of the soul," in which she represents an agate lying in the earth, gaining the sole idea of its existence from the roots of a violet which embraced it. The knowledge that there was such a thing as life filled the stone with so intense a desire for an existence having an element of vitality—life—that it wished itself away into a lichen that grew up in the sunshine and felt the inspiration of the winds of heaven. But that aspiration which belongs to spirit, and seems even to impel upwards insensate matter, stirred in the new existence of that spark of spiritual life, and as it learned of higher forms of being—higher grades and developments in life, again it passed upwards into a higher sphere, and again, and yet again, until through varied shapes it passed through a nightingale into the musician Mendelssohn.

This poetical fiction was once accepted by a school of philosophers, who certainly had more foundation for their theory than many another school and sect, which have obtained more widely in the world than this, for though these men took the shadow for the substance, still there *was a substance* to cast the shadow, while many a school of sectaries have built their faith on the "shadow of a shade."

But to return from this speculative soul-making to the passing of the mineral up into the vegetable kingdom, and there trace the footprints of nature's upward pilgrimage, from the first initial Algæ of Geologic revelation, to the elaborately perfected Phænogamia of the historic period. We find in those vast fossil forests revealed to us in coal beds, mostly gigantic Cryptogamia, luxuriant and wonderful, yet lacking many of the organs common to all the more perfectly organized plants, of our present more diminutive, but more numerous, and more highly finished genera of plants.

But beyond this passing from one species of an earlier period to a more perfect one of a later, there is this perfecting process going on in the different varieties of the same species, under favorable circumstances many varieties being coined at will by the ingenuity and skill of the human race. Thus a single variety of wild rose, perfect in itself, in one sense of the word perfect, can be varied by climate, and soil, and cultivation, into innumerable phases of beauty, "differing one from the other as the stars differ in glory," yet all developments of the ideal perfection, which, though absolute and infinite, admits of more variation than imperfection, which is comparative and limited.

Naturalists tell us that double flowers are monsters; but the universal consent of all tastes, civilized and savage, has pronounced this development of educated nature to be beautiful, and thus contradicts the idea of deformity which is always hideous.

The wild crab tree was perfect as a crab, but not as an apple, until its wild bitterness was won away by the faithful and kind education given it by those it so plenteously rewarded with the luscious richness of its harvest gifts.

Some of the higher orders of plants seem to be endowed, not only with sensation, but with instinct, discovering by some unknown power, the presence of various substances not in actual contact with them. Thus several species of vines, such as pumpkins and melons, will in one night turn the straightforward direction of their growth to reach a vessel of water placed a few inches to one side of their path. Twining plants will reach a support placed near them, and tendrils remain uncurled, stretching out their long fingers to find something they can cling to, which, when found, they lose no time in grasping.

The Dionea sets snares for the luckless flies it lures with its sweetness, and then, like a spider or a coquette, imprisons and pierces its victims with its thorns; and these insects are said to supply an important item of food to this carnivorous vegetable. The pitcher plant in the desert, opens and holds up its cruse to catch the rain and dew, and then shuts down its lid to keep the sun from stealing away the treasure. The mimosa, most like an animal of all, shrinks from a touch, and even dies to escape the cruel handling of those who wish to gratify their curiosity at the expense of its sensitive organism.

All these approaches towards sensation and volition, betoken advancement towards the object of nature's involuntary yet incessant search—*a something higher*. But as we seldom find the realization of our expectations in the place or manner we had calculated on, so if we look among these higher developments of plant life for the place where the vegetable kingdom is to be found in nearest approximation towards the animal, we shall not find it. The bottom of the valley of humiliation is the nearest place to Heaven, so among the lower Algæ, the very alpha of vegetation, we are to look for those strange animal plants, or vegetable animals, which link the two kingdoms together and make their boundary lines indefinite. From these the transition is rapid to the zoophytes, and up, through an almost endless chain of marine and terrestrial animals, whose cold, and often colorless, blood seems as lifeless as the sap of trees, or the crimson juice of the bleeding sanguinaria.

But the organism that approaches nearer and nearer the perfect, not only through all the successive Geologic periods, but up from one race to another of existing animals, develops more and more of that species of intelligence called instinct, until the more fervent action of the vital forces brings warmth, and dawning glimpses of intellect, which, growing finer and purer, less gross and more flexible, becomes complicated—active with the impulses of will, and the teachings of something very like reason, if not its very self, until it emerges from this long, long line of successive, ever-advancing existences, into the highest type of animal existences yet known—the Human.

When we come to the more perfectly organized plants and animals we find a new and important feature developing itself, and this is the more *distinct individuality* of the single members of any species. Hitherto we have been considering differences shaded off into almost indistinguishable likenesses, and diversities becoming apparently identical. From considering these seeming transitions, and likenesses between differences, we turn to scanning those yet stranger individual distinctions which separate the universe into items which must for ever remain distinct. Among many plants, copies of original varieties can only be obtained by grafting, budding, cuttings, &c., the *seed* producing new individual varieties endlessly; but when we come to animals, this exceptional mode of producing synonyms no longer exists, and we can find only individuals, in a great measure resembling each other, it is true, but in the superior races, so essentially differing in character, as to require separate study to understand the treatment of each individual, in place of the generalizing process which sufficed with plants arid the lower animals.

This individuality, in some degree, exists among all organized beings, and in each individual can be clearly, definitely, and easily traced, the advancing steps of nature towards perfection. In plants—even in the simplest and lowest, the individual can be traced from the minute spore or seed, through the various stages of germination and growth—its first singular leaves, and after growth of foliage, its blossoming and fruitage, and until the ascending circle is completed in the seed containing the life of another generation.

The lower animals, like plants, change more externally than those of a higher order.

Thus the mosquito passes from its egg-built life-boat into a water monster of minute dimensions, and afterwards leaving the skin which protected its aquatic existence as a legacy to its native puddle, it creeps into upper air, furnished with wings and a song. Numerous reptiles perform this pilgrimage through two elements in order to perfect their natural developments, and the whole or nearly the whole tribe of winged insects undergo the change from creeping to flying things. Some of these transformations are so beautiful as to be assumed as types of mankind's angel life, following the human one.

The Greeks carved butterflies on their tombstones as emblems of the changed condition of the individual whose forsaken shell lay under the monument.

The larvæ, emerging from the egg's prison shell, and crawling, dim-sighted and slow, as it grew to the full size and perfection of its worm-life, feels the hour approaching when the caterpillar must die and the butterfly be born; so, sometimes winding itself in a silken shroud, and sometimes casing itself in a hard horny shell, it falls into a sleep from which it awakes not a blind, trailing worm, but a new creature, many-eyed and rainbow-winged, no longer doomed to gnaw its daily food from bitter leaves, but to float ever among flowers, and drink their full cups of sweetness.

Of all the emblems of the myth-loving Greeks, I know of none so beautiful as this type of growth from the low and loathsome to the perfected and beautiful individual creature.

From the bare nestling, with its great head, and sealed eyes, and its wide gaping mouth, through the chirping fledgling, to the full plumed singing bird, as from the babe through childhood and youth to the full grown adult, we see always before us the operation of the same universal law of progress, the same tendency in the natural to perfect itself.

Thus far have we traced the material, visible, tangible revelations of Nature, and their unvarying teachings point us forward and upward. She has qualities which we are too little and too ignorant to weigh or measure, properties as undeniable as they are incomprehensible and inscrutable. But no one of them contradicts this one great lesson; though we do not know positively that they have been subject to this progressive law, neither do we know the contrary.

Light may be precisely as it has always been since its creation, but not *where* it has always been, for it is ever moving, ever fleeing through space, like a thought, to reach some more distant spot in the universe.

Heat, too, may be an eternal element, changeless and perfect from the beginning; but if electricity has, we have no evidence of the fact, for its phenomena have progressed essentially since the scientific observations of men were first recorded. It is not yet three hundred years, I believe, since the Aurora Borealis, generally believed to be an electric phenomenon, was first observed. If it had existed prior to that time, as it now exists, people could not fail to observe it, and a people superstitious enough to be frightened out of their wits by a comet, would surely have as well recorded their terror at a sky all ablaze with strange cold flames, such as light some of our winter nights, as to have written down their exorcisms of Satan in order to light the darkness of an eclipse.

History and analogy, then, go to teach us that universal nature is, and ever has been, climbing upwards towards perfection. True, all organized being is reducible to a few simple elements, in which there is no such law apparent; yet these elements seem restlessly seeking to enter new combinations, and in these combinations to become part of new organizations, and there obey this organic law of the universe.

In following the material creation up from its inauguration, we come to places where were successively developed animation, volition, will, intelligence, and reason, all of which, except the first, find their highest development in the Human race. All the passions, feelings, and most of the instincts of the lower animals, are heightened, spiritualized, and intensified, and form an essential part of the invisible man; and these faculties, or rather feelings, seem, when we find them in their less perverted and more natural state, like all lesser things, to recognise the existence of a higher life, and by their endeavor to slough off their unnatural corruption and depravity, seek to assimilate themselves to their ideal type of natural perfection.

All the intelligence and reason of the lower animals find a deeper, stronger, more elevated, and very much more extensive and perfect development in the intellect of man, and this superiority of a force, stronger than the physical, is acknowledged by the lower races. The hyena and tiger quail before the steady gaze of a soullighted eye, and the hungriest bear will not devour the child who fixes its eye-balls on his own.

But it is not the supremacy of the intellect alone which gives the human will its power over the brute. In him there is made manifest a new faculty or force, not demonstrated to exist in any other animal the beginning, the initial, the alpha, the algæ and zoophyte of a new ascending series of being—winding its spiral circles above him, through angel and archangel, seraph and cherub, up to the All-Beautiful, All-Wise, All-Good, All-Powerful, Essential and Eternal Spirit of the universe.

The existence of this ascending series of being, beyond the point where it sojourns in the flesh, because inaudible, invisible, and intangible to the senses, has been doubted by some who believed that nature was perfected in man— that he was the ultimate, the highest—and that the perfection of the natural was limited to man's capacity for improvement and perfection. But this class of persons has always been, and probably always will be, few in number and limited in influence, for in man, as in all the rest of nature, only more self-conscious and positive, heave and throb those irresistible "propulsions from the night," which never move idly towards nothing, but always directly in obedience to the strong upward attraction of higher forces.

Does the restless tide disturb the great deep in idle wantonness, or because the resistless attraction of an upper world woos it to follow its aspirations? Why then, think ye, should the great tide-wave of human aspirations after the higher and holier existences, be called the aimless heaving of native unrest?

This restless, far-reaching instinct of the soul aspiring towards its ideal, is our best and highest proof of the existence, and the best demonstration of the nature of the unknown world to which we are hastening, and of the beings inhabiting it, as well as of our own immortality.

That God made man in His own image, we all prove by constantly inverting the process, and making Him in the image of our inner selves. Every sort of man makes his ideal deity something like himself—fashions him according to his own ideas of perfection; so we find the general ideas of God exceedingly diversified and contradictory, as well as indefinite and confused. The man only a little above the beasts has a strong, passionate, wilful Deity as his mind's ideal—the cruel man has his Vengeance and Terror God, and the beast man worships his crocodiles. Adam, in his ignorance, imagined the Creator to be as gullible as himself, and sought to hide from his presence. Some philosophers and many poets, supposing their faculties to have each their prototype in a separate divine existence, based on man's multifold nature strange systems of polytheism. Others supposed all the sinless universe to be pervaded by an essence they called God. But the sensual represented their god as a devil; and the philosophers described a "keen, cold" matchless intellect as their deity; a set of weaker heads and warmer hearts learned God's diviner attributes of justice, mercy, and love. Afterwards—ascending ever—prophets with both heads and hearts, learned and taught how the manifold manifestations of God were only parts of a whole—how the all-pervading Life Essence, and the guiding Intelligence, and the Spirit of love and compassion, were all but parts of the Great First Cause.

Slowly and vaguely, dimly and indistinctly dawned these successive ideas on the world, and came at last to assume a character for Deity which was the immutable and absolute right—the abstract Spirit and concrete Essence of perfection—the ultimatum of the universe. But men differ so widely in their notions of right that their ideas of God are as diversified as ever, for there are those of every grade of worshippers known in any age, still living and worshipping their highest idea of the good and true, from the votaries of serpent gods to the disciples of the Universal Heart.

But God, not being limited by man's base ideas of him, has so ordained His laws, or they so exist in Him as a part of himself, that they can nowhere be transgressed with impunity, thus fixing limits and boundaries to right and wrong, making truth an absolute and positive thing—a real, demonstrable fact in existence—so that man has but to read nature's open scroll, to learn the truths it behoves him to know.

But nature's scroll is a sealed book to most eyes—a chart of life's voyage made out in unintelligible hieroglyphics, which men are too lazy or too stupid to learn, hence the blind groping of the race for ages to find the way of truth, to which every tree pointed with innumerable fingers, of which every brook babbled, and every breeze whispered—the thunder pealed out the lesson, and the birds warbled it everywhere—the mute flowers breathed it out in odorous sighs, and the heart of man echoed the voices, yet lost the meaning of his teachers.

Other revelations were given, but the burning truths that came from the inspired lips of God's chosen, fell on the same dull ears and clouded understandings, till their divine harmonies were tortured into discord, and men learned but disjointed scraps of the great truth.

But as "Chaos and old night" faded and dissolved before the light and life of creation, so pass away these dark and cloudy troubles from the human soul. Slowly do the dull ears and duller understandings of men come to hear and understand the demonstration of truth's absolute law, and more slowly still do they come to yield it obedience,—*but they are moving!*

But heavily and slowly, like blind Orion climbing up the mountain, bearing his guide up the steep to the summit, where he might meet the first rays of the morning's sunlight, which were to unseal his closed lids, so climbs the great world, burdened and blind, towards the all-revealing future. The past clings like tendrilled vines, or thorny and hooked-fingered brambles, to her wet trailing garments, and holds and hinders her from moving on, since she knows not the pathway, and her guides are almost, or quite, as blind as she.

In the many-voiced chorus of the by-gone, she seldom heeds or hears the tones of prophecy and "lofty cheer," but listens, with flagging steps and weary heart, to the oft-recurring refrain, "Tempt not the untried ill in search of the unknown good—keep what thou hast—it is thine"—and were it not for the restless life-growth in all things, the world, obedient to its conservative teachers, would stagnate and die.

This world-growth may be likened unto a handful of acorns cast on the earth, and one of them felt the sunshine and the dew, and received them as coming from Heaven, and her heart swelled within her till the smooth hard prison-house that held it, burst apart with the power of the life that was in it. And then the conservative acorns around it began to cry out, "O foolish acorn! when Heaven gave thee so smooth and beautiful a shell to protect thee from the cold rain and withering sunshine, why wilt thou ungratefully despise its gifts, and choose these restless elements for thy companions, when they will certainly destroy thee? A while ago thou wast an acorn, but now thou art nothing but a misshapen mass of fermenting material."

But the acorn felt the inspiration of life, and answered, as inspired enthusiasts always answer, in a language her sisters could not understand—and they chid her, as one gone mad.

But after a time, she stretched out her hands so far to embrace the free air, that her two first leaves, all wrinkled and ridged, spread out in the sunshine, and her feet clung closely to the moist earth, and she stood, a living, growing thing.

And her sisters said, "Who would think that nondescript, bold green monster, staring in the face of the sun, had ever been a beautiful white acorn, modestly folded in her brown house? But she will soon reap the reward of her temerity, for one baby footstep would crush her to pieces now, while we are as safe as ever."

But the young plant felt her soul swell with more life, and as she looked up to the quiet stars beckoning above, her heart opened, and began to put forth a new leaf—thinner, and wider, and deeper colored than the first; and she rejoiced—for growth was so happy, and so glorious! But when her sleepy sisters awoke in the morning, and saw her again expanded and changed, they cried out the more exceedingly against her, for her fickleness, and told her to behold how constant *they* were to first principles and old truths. But as the plant grew, neither by creed nor theory, nor even by reason, but by instinct, and inspiration, and necessity, she was "grieved in her heart," and having no cavilling sophistry wherewith to answer their stereotyped logic, and the living words of truth "being as foolishness" in their ears, she grew on and "answered not a word."

But when the Summer was ended and Autumn came and painted the green leaf red, and took away the first pair, the sisters said, "Behold! she is dying! The frost cuts through her flimsy foliage, and destroys her, but *we* are snug and safe where nature put us, and defy the storms. Let us take warning by her folly, and be contended with our condition." But the young oak cried out earnestly, that nature meant she should grow, or she would not have given her the power, and that she should not die but live. And they said, "Then was it intended that man should sin, or he would not have had the ability!" and they laughed her to scorn.

Then the snow came, and they slept.

But when the "Winter was over and gone," and Spring filled the oak with fresh vigor, it branched and spread, and grew and thrived luxuriantly, and the fault-finding acorns grew silent, or only grumbled in whispers; but afterwards when the oak found honor, and the acorns were unnoticed, they began to envy their sister, and take pride in her, and try to imitate her, but decay had fixed upon them, and worms gnawed into them, and they learned, too late, that when nature bestowed sunshine and rain, she meant they should be used, as well as her temporary sheltering gift of shells.

So the human part of the natural world, like the life-inspired acorn, shall grow in spite of sneers and cavil—in spite of frost and storm, of time and delay—and become beautiful and glorious, even as the oak became monarch of the forest.

Men have already begun to listen for the true teachings of the great Past, which stands, sybil-like, chanting her oracle and prophecies, and pointing to the records of her progressive march, graven on all the rocks adown the stream of time.

To those who have learned to hear aright, her song teaches us to struggle upwards towards the future, as she struggled to reach the present, ever nearing that great object of her search and ours—the highest ideal—the truest truth—the ultimate perfection.

The poet tells us that

> "There is a mighty chain of being
> Lessening down, from Infinite perfection
> To the brink of dreary nothing"—

and it is true. The converse of it is also true, and we are tending upwards in the ascending scale, enlarging and growing, from the atom up to Deity, and when the struggle shall cease, and that harmonious march up towards the infinite—that march which is to be trodden through eternity, to the music of the spheres, shall be begun, then shall each item of the universe in the perfection of its own nature

> "Be one with God, and God is All in All."

A PILGRIM PAGAN.

THE dusky twilight of all time had come,
Earth's long and stormy day was closing fast,
And night, not earth-born like all other nights,
But born of space as stars are, swallowed up
The lingering relics of the by-gone light.
The Norsemen's fabled Raggnarokk had come,
When all the Gods of earth were to be judged
And swept away, and only "He whose name
No man dare utter," should be Lord of all.
And he, the fiery king of endless life,
Judged the pale spectres of the waning world.

"Come forth!" said the Eternal: from the tomb
The long dead Ages started back to life,
And bursting through the mould and dust of death,
Rehearsed the deeds of every by-gone day.

Thus witnessed one before that awful court:

There have been Angels, tho' in various guise,
From the beginning through the lapse of time;
Each martyr'd in his turn a thousand ways,
Each burthened with a host of evil names,
And scourged, and mocked, and cast aside in scorn,
Trampled and hated; yet 'was marvel strange
That while the Angels thus were bufieted,
They each had altars whereon incense burned,
And they, the buffeters, sung psalms of praise,
And offered incense to the names of those
Whose incarnations they despised and slew.

The Angel, Truth, had temples old and new,
And throngs of worshippers bowed to her name;
And yet the very temple's selves—their pomp

Of marble colonnade, and fretted roof,
Gold-gleaming spire, rare woods of distant climes,
And costly deckings—were but wooden cheats,
Poor, painted make-believes, whose honest cracks
Showed the plain pine beneath the rosewood lie
And rich mahogany falsehood, deftly told
By cunning workmen and their pots of paint.

This shabby showroom of men's vanities,
In solemn mockery named the House of Truth,
Was filled to flooding with a motley throng,
Who went—because it was the day to go—
Because it was the custom of the place—
Because the bell rung, and their clothes were on,
And folks would notice if they didn't go,
And folks would see their finery if they did.

Rolled thro' the vaulted roof the organ's peal,
And the sweet chant of voices in the choir:
The hymn once uttered by some earnest soul
Enhedged in darkness, seeking for the light,
Was chanted parrot-like by choral bands,
Who lost the meaning in the silv'ry sound.
Thus ran the burden of the solemn song:

> All art Thou, but we are nothing,
> Help us, Oh! thou Truth sublime!
> We are false, but Thou art faithful,
> Teach us all Thyself divine!

> Wait we mourning round thy altars,
> For the life that Thou dost give,
> Take our broken hearts and heal them!
> All our worthless gifts receive!

Followed the prayer in measured accents doled,
And sermon, tedious as an oft-told tale,
And then a pause. In thro' the outer door,
There passed a being most serene and calm,
And stood before the altar in the aisle,
And thus he spoke: —"Oh! fellow worshippers!

I come from a far realm away to join
The congregation of the friends of Truth,
Who has no temples in the land I left.

"There is within me that which turns from wrong,
Which shrinks from treach'rous deeds and lying lips,
An outgrowth unto all sincerity—
And with such aspirations as shall grow
To deeds, my spirit searches for the Truth—
The infinite—unfailing—utmost truth!
Enrol my name among the names of those
Who seek in her the ever highest good."

He paused, and ladies smiled behind their fans,
To think that such a funny looking man,
With such queer clothes as this strange pilgrim wore,
Should make a speech like that, so out of place,
When no one asked him, and none other spoke.
Could he not wait till the appointed time
For offering names? And then he could have made
A speech, and worn a garb, more à-la-mode.

The men drew down the corners of their mouths,
And looked half sneering towards their shining boots,
And wondered what the stranger's business was,
That would allow him to afford to be
So very much in love with thriftless Truth.

A few, more zealous than the rest, were glad—
Some that the neighboring temples all would gaze
With envious admiration on this new
Converted heathen, that none other had,
But was a unique feather in their cap.
Some liked his eloquence, and some were pleased
To hear their language with strange accent mouthed
In voice so musical by mouth so fine;
And some were glad that their monotonous lives
Had had at last an interesting break,
And their stale gossip a fresh flavoring spice
For the next week at least. But there were some,
"A beggarly account" of "nobodies,"

'Tis true, yet still a living few, whose hearts
Were thrilled with deeper joy than tongue may speak,
That there was one in all the earth to dare
Exile and ridicule for Truth's dear sake.

Then spoke the minister: "If you have read
Our Articles of Faith, and can confess
The Creed and Catechism without flaw,
After initial rites, if none dissent,
We will admit you to our loving band."

The half-converted pagan looked aghast!
"I never heard of all the things you name,
Nor can conceive of an initial rite
Would make me love Truth more than I do now!"

"'Tis not for that," replied the minister,
"'Tis but a form of setting you apart
From all the world that does not follow Truth,
I will repeat the creed to you, and if
You can confess it afterward, 'tis done,
Say, 'I believe in Truth, the pure, the whole,
That she hath many eyes that she may see
All that exists; some scanning up, some down,
And some the limitless expanse between,
And hath as many tongues to speak of all
That she beholdeth in the universe,'
Can you repeat thus much?"

                    "And why should I?
For I know naught of this, and how believe?
It matters nothing to my love, if Truth
Hath but one eye, or two, or scores of them,
So that she still be Truth. Why should I care?
She may be many tongued, for aught I know;
I would she had ten thousand living tongues
To speak in all the earth, and thus cry down
The clamorous lies with which the welkin rings.
I would exult in such triumphal song,
But I will listen with love's keenest ear,

In every tempest lull of falsehood's storm,
For the deep whispers of her still, small voice,
When she is bidden not to speak at all;
And when men make it crime to speak her words,
Then will I speak them faithfully as now—
Will wear her badge when 'tis a badge of shame
And honor her thro' sneers and bitter scorn,
And tho' I have not numbered all her tongues,
I love and I believe in utmost Truth—
Pass by the rest, I cannot say it now!"

"But 'tis essential that you should *believe*,
And if 'tis ignorance that makes you halt
Become enlightened 'ere you seek to pass
The sacred pale of our especial Truth,
Outside of which all other Truths are vain!"

"I know," the heathen said, "I must have all,
That one great central part of Truth withdrawn
Will make all others seem but partial truths,
And half-truths are part falsehoods at the best,
I want the whole, what is entire, alive,
Breathing and throbbing, quick and warm with life,
A dead Truth is as worthless as a lie;
If I take verity at second hand,
And say 'tis so, because *you* say 'tis so,
What is it better than dead truth to me?
A falsehood were as quickening to the soul!"

"But you want Faith to guide where knowledge fails
We must believe those things we cannot know."

Slowly the puzzled heathen turned away,
And stood alone beneath the endless sky,
Baffled of human aid lie must toil on,
And slowly proving all things, hold the good,
The good alone, against the faithless world.
The garb those sneering saints had thought so odd
Around his great, true heart he wrapped, and thus

Unto his inmost soul the hero spoke:

"Without conditions will I come to thee,
Thou all pervadmg Soul! The cumb'rous creed—
The cramping rite—the mouthing of strange words—
I leave for those who find a use for them,
But Thou, whose speech the soul can understand,
In the dead silence of the voiceless night,
Thy words come to my feverish soul as rain
Comes dropping on the dust-encumbered grass,
And freshens all its wilting stems and leaves
"With its cool message from the far off clouds.
My soul is drooping 'neath its gathered dust—
Drop down thy starry dews and quench its thirst,
Oh Truth divine! and make me strong
To seek Thee evermore—and seek *alone!*"

The congregation in a motley stream
Had poured out past him with averted eyes
And left him as a harmless half-crazed man
A stranger and alone among the tombs.

The witness ages vanished from the stand,
But, by the throne of Him who judged the world
Calm stood the steadfast seeker of the Truth
As one whose utmost toil was lost in peace.

A DREAM ANTHEM.

PRIEST.

BELIEVE ye in the Lord of Hosts, the Ruler of the sky,
The universal Mother heart, the Lowliest and Most High?
Believe ye Him, the Soul of all, who wears eternal space,
And lives, the moving vital thought of each created place?

CONGREGATION.

Aye! Aye! believe we in the Lord! and worship from afar
The King upon his great white throne above the farthest star!
We build up temples to his name, which pierce the very sky,
And shout our praises loud and long that he may heed our cry.

PRIEST.

Believe ye in life's holy Truths, whate'er may be their name;
Tho' scorned amid earth's chase for wealth, for pleasure or for fame?
Believe ye in the soul of Love, which claims each living thing,
However low, disgraced, or mean, the source from which they spring?

CONGREGATION.

Aye! we believe in lofty Truth! Truth is great and nobly born!
But bow we *not* to thing so mean as to be sport for scorn!
And we believe in Love divine, all conquering and sweet,
But waste it not on outcast ones in the prison and the street.

PRIEST.

Well speak ye of twin Love and Truth! right nobly born are they!
The right hand and the left of Him, the king to whom ye pray!
Eternal as our Sovereign Lord—the Lord of Heaven's high host,
The outcast Love! the slighted Truth! God's Son and Holy Ghost!

Look not above the farthest star! God whispers in the air,
And instinct with his breathing life glows nature every where.

The time will come! The time will come, when on every stick and stone
The blindest eye can but behold the Eternal's living throne!

On all the doors, on all the walls, from the threshold to the eaves,
On all the buds and bursting flowers, ripe fruits and trembling leaves,
And on all things living shall ye see, as written on a scroll.
In characters of light glow forth the universal soul!

Believe ye in the Lord of Hosts?
(CONGREGATION.) Aye! believe we as before!
(PRIEST.) Believe ye that all Truth is God?
(CONGREGATION.) Aye! and worship evermore!

ALL.

We accept Thee, God of Mercy! Love, and Truth, and Life art thou!
Accept us while we bow before Thee! To no other do we bow.

THE LOST LAKE.

Near Smethport, in Pennsylvania, there is a great pile of conglomerate rocks, known by the name of the Devil's Den. During a pic-nic at the place once, a gentleman who demurred at the popular mode of devoting all grandeur to "the gentleman in black," proposed taking the first initial of each lady's name then present, and coining a name for the place. The name was Eel-e-a-ha-o-aa see, so suggestively *Indiany* in sound, that the following commemorative poem grew out of it ever so many years ago:—

PART FIRST.

NATURE in wayward mood had thrown
O'er yon mountain's top a giant stone;
The river's waves no longer rushed—
The wild bird's warblings were hushed—
The woods grew dark with sickening fear
 Its heart-pulse faint and cold,
As downward on their wild career
 The broken fragments rolled.
There in rough disorder piled,
O'er that mountain wide and wild,
'Mid tangled laurels, dark and rude,
The rocks grew grey in solitude.
The loosened earth slid slowly down,
Leaving the hill-side bare and brown.
It left the mountain's hoary head,
And rested in the valley's bed;

The river's waters sought in vain
To trace their pebbly track again.
Back to their fountains, dark and cold,
The imprisoned waters slowly rolled;
'Till upward 'gainst the opposing heap
 They wider, mightier, rose;

Then o'er its brink they wildly leap,
   Free from their dull repose.
The river's dried, deserted track
Welcomed the fresh, pure waters back.

High on the mountain's dusky brow,
Defying wintry storms and snow,
Three walls of solid stone uprose,
And o'er their massive, grim repose,
Rested a roof of one great stone,
With creeping evergreens o'ergrown.
By the low doorway on the east
Sat that lone mountain's hermit priest.
Near, thro' mossy crevice, crept
Waters where the lichens slept;
Drop by drop those waters fell
In a moss-enamelled well,
Near that grey old hermit's cell.

Across the lake, and on its brim,
In the forest's shadow dim,
A wigwam raised its lowly head
Close beside the water's bed,
And in that hut a warrior bold
In prosperous peace was growing old.
His sons were "swiftest in the race,"
Their bows were surest in the chase;

His daughters, with fond mothers' pride,
Their children's tottering footsteps guide,
Save one—the flower of vale or hill,
Who watched, and loved, and cheered him still.
Evening was rosy in the west,
And tipped with gold each mountain's crest,
When a tall form, as oft before,
Darkened that lowly wigwam door;
Strong was his soul and firm his tread,
Ere the low doorway bowed his head;—
He spoke; Nayugeh could not brook
The lightning of the chieftain's look.

"Old Eagle of the pine clad hills!
    A day of blood is breaking!
Sleep no more by purling rills,
    Thy warriors are waking.
Pale children of the rising day
Come like wolves upon their prey;
Like wolves upon the wild deer's track!
Like panthers we will send them back!
To-morrow is our council held."
Then forth he hied—the war-whoop yelled—
And to some other brave's abode
The Cougar of the mountain strode.

The council met, then from the earth
The red war-hatchet sprang to birth;
The war-pipe's smoke 'tween the old oak limbs
Disturbed the wood-wren's matin hymns;
And then arose to the peaceful skies
The smoke of the white dog sacrifice,

And to that sacrificial feast
Came Eel-e-a-ha-o-aa-see's priest,
From the lonely-mountain side,
Where none but he might dare abide,
Where naught polluting e'er might dwell,
From the moss-grown hermit's cell,
Where he his sacred treasure kept,
Guarding while he walked or slept,
That treasure that he deemed divine,
The sack of holy medicine;
Like the seraph-guarded ark,
    Of which Scripture's pages tell;
Like a strayed and stranger spark,
    From the fires of Israel!

'Twas over! Then from south and north,
From east and west went warriors forth;
Armed with war-club, spear, and bow,
    With tomahawk and knife;

Strong hands were there to deal each blow!
   Strong hearts for blood and strife!
Then arose the war-whoop's fearful yell,
Wide over plain, and hill, and dell,
And echo came back from hill and glen,
Like the yell of demons instead of men;
And echo caught up the hideous cry,
And bore it aloft toward the burning sky.
Then silence o'er the valley crept,
As if all things living slept.
The wild birds' morning songs were o'er,
The hum of man was heard no more,—
Not a breath of wind there stirred,

Not the faintest sound was heard,
Save where the lake leaped from its thrall,
Dashed the noisy waterfall;
But the leaves its babbling drowned
Into a sadder, softer sound.
The smoky air wrapped every hill,
And seemed to hush its waters still;
But the silence was too deep,
Too unnatural for sleep.

In Eel-e-a-ha-o-aa-see's cell,
The priest was chanting o'er his spell,
While watching, more in awe than fear,
Stood the father of the youthful Deer,
While she, Nayugeh, stood afar,
And prayed to the God of peace and war,
To send her Father safely back
From the war-path's dangerous track.
Then in her heart a stifled prayer,
That to breathe she did not dare,
For the Cougar that stood gazing on,
The Hermit and his myrmidon.

The Hermit watched his magic fire,
Whose blue flame flickered pale and dire;
As he muttered his spells in a tongue unknown
To a fiercer blaze was his watch-fire blown;

Quickly the red flame leaped on high,
With a glory that paled the burning sky,
And eagerly seizing a pine tree's stem,
It crowned its cone with a diadem;
With its wreaths of smoke and its tongues of flame,

A terrific torch that tree became.
Then sudden the yell of the Hermit rang,
Till the echoing hills caught up the clang—
Then sunk his voice to a muttering tone—
Then chanted as if to himself alone.

"Not till yon lake shall cease to spread,
Over the quiet valley's bed,
Not till the weeds that in it stand,
Flourish upon the naked land—
Not till its barrier is strewn,
Where Nonandah's limpid waters moan,
Not till yon lake deserts the glen,
Shall we yield to a race of paler men—
When the Lake of the Forest no more shall be,
Our tribe shall die like the foam of the sea.!"

He ceased his low, complaining wail,
The warriors sought the foemen's trail—
Nayugeh sought with wondering gaze,
The pine tree torch's brilliant blaze,
Which stood aloft like a beacon light,
Lighting the shades of coming night,—

"Why follows the Deer on the Cougar's track?
Will the Wild Doe welcome the Panther back?"
Nayugeh turned with a startled sigh
To meet the dark inquiring eye
Of a sister beloved who was standing nigh.
"Come, sister, my lonely wigwam share—
'Tis gloomy unless the loved are there,
The Eagle has flown toward the morning sky,
There are none to comfort or keep thee nigh,
But we'll gather herbs from the solemn wood,
And draw the fish from the silver flood,

And watch and wait till our warriors come
Together to welcome the wearied home."

PART SECOND.

"I came, O Wild Doe of the hills!
Back to my native rocks and rills,
With my life blood ebbing slow away,
From a wound whose bleeding I cannot stay.
Ye cannot follow on our track,
And I may never more come back;
Where canst thou flee in the hour of wrath,
When the Eagle has fallen in war's red path,
When the Cougar is far from Nayugeh's call,
Will the child of the forest die in thrall?"

Nayugeh had watched his kindling eye—
Her bosom was burning to make reply;
But a daughter's love and a maiden's fears
Burst forth in a passionate gush of tears;
From her sobbing lips no answer came,
Save a murmuring of her father's name.

"Cold on the far off plain of death,
Sighed forth thy father's dying breath;
And I alone am left to tell,
Where the Eagle of the mountains fell.
No pale face bore his scalp away,
He lies no colder there than they."

"Thou hast avenged him then!" she said,
Raising her lone and sorrowing head,
"But hast come alone from that dreadful fray?
My Warrior brothers, where are they?"

"Cold on the battle field they lie!"
Was the mountain Cougar's sad reply,
"Our shattered tribe are gathering,
Weak like the flow of a sun-dried spring;
But the pale men came like a river's tide,

When the snow melts off the brown hill side;
Whither, oh! Doe of the forest dark!
Wilt guide thy canoe of the birchen bark?
Where, oh! maiden mild and meek!
Hide where the pale men will not seek,
For their hounds can follow thy footsteps well?"
Nayugeh murmured, "I cannot tell!"

"Follow me, Nayugeh, in this mountain path,
What though the storm God growl in wrath?
Is the eyrie rocked eaglet of tempests afraid!"
"No!" said the voice of the following maid.

On, dark rolled the river, its wild rising flood,
Fierce howled the storm through the loud wailing wood,
Yet up through the shrieking tempest's wrath,
They follow that tree strewn mountain path,
Up, past the Hermit's empty cell,
Up, where the trees wild crashing fell—
Up, where each grim rock raised its head,
Where Eel-e-a-ha-o-aa-see spread

Its platform dark of giant stones,
Above the wrung trees' crashing groans,
Here the Mountain Cougar climbed at last,
And paused—his clambering was past,
At length Nayugeh near him drew,
Her weary march was ended too.

A ragged aisle the wind had made
Down before them to the glade.
The pent lake's waters rose up still,
Chafing the foot of the rocky hill.
Ah! never before since those cabins stood,
In the shadowy marge of the wide-spread wood,
Had risen so high that river's flood—
When the mountains shook and the valleys groan,
With an echoing crash in thunder tone—
The barrier burst! The Lake was gone!—
And the cabins that stood on the outlet's shore,
Were lost in the waves and seen no more.

"We are the last!" the Cougar said,
"The lake is gone—our tribe is dead—
Ere night shall darken o'er the sky,
The Cougar of the Hills must die,
And one lone Doe in the pale man's path,
Is all that is left to stem his wrath—"

"Oh better, far better to die with thee!"
Responded the wild Doe earnestly.
How glow'd o'er the face of the dying chief,
The light of a smile as bright as brief!
Then answered he with a deep drawn sigh,
"I brought thee here with me to die!
O! deeply these rifted rocks between,
With a feathery fern and a laurel screen,
To whisper with winds that are floating by,
'Twere sweet for the last of the tribe to lie—
'Twere sweet for the Panther and Doe to die!"

The moss grown rocks the notes prolong,
As they sang together their funeral song.

Howl! tempest in the forest, howl!
    One tribe no more shall heed or hear!
Above them shall the wild wolf prowl,
    And undisturbed pursue the deer.

Weak are the hands once strong in battle,—
    Cold are the hearts once warming our home,—
The bones of our braves the foxes shall rattle,
    Over each heart's deserted stone.

We come to Thee, great ruling Spirit!
    To thy happiest hunting ground—
The land that the good and the brave inherit,
    That the pale face has not found!

They ceased, the Cougar's failing arm,
    Was round the form he loved the best,
He seemed to shelter her from harm—
    He looked as if he but caressed.

Then kneeling on the giddy brink,
    He sheathed his knife within her heart—
Then in his own. The mosses drink
    The blood of those who would not part.

O! deeply those rifted rocks between,
With a feathery fern and a laurel screen,
To whisper with winds that are floating by,
The last of the tribe of the Eagle lie.
The Lake of the Forest no more shall be,
And its tribe are gone like the foam of the sea.

"LOVE IN A COTTAGE."

PART I. THE PRELUDGE.

Orthodox according to Romances; Apocryphal according to Fact.

NEAR a "dark rolling river" by "willows o'erhung,"
   Stood a "lowly white cottage" with shutters of green,
Where the creeping vines round the windows clung,
And the "gnarled old oak" its shadow flung,
   A broad armed and beautiful screen;
Round one room the flowers were growing
   "With a mute caress,"
As tho' thro' shadowy foliage showing.
  "The inmate's loveliness."
For "lovely as the morning" smiled
That "sweet and lovely" "forest child."
Her eyes were "liquid hazel," beaming
  'Neath the "dark fringe of each lid,"
Her "alabaster neck" was gleaming,
   Half by its "curling drapery hid."
Her beauty might bewitch the sight
Of any frigid anchorite;
   But the "soul that beamed within"
   Was something deeper than the skin!

Young Oscar from the mill one day,
Chanced to be wandering by that way,
When to milk the cows, thro' the river's slime
Waded the barefooted Angeline.
They met beneath the spreading tree,
And "their first glance" "sealed their destiny!"
While seated on the milking stool,

Beneath the willow's shade so cool,
The maiden sang a song more sweet
Than songs which summer evening greet.
A joyous thrill through each nerve sprang
As the music tone through his young heart rang—
Full many a glance she sidelong cast,
Ere those hours on golden wings had passed.
His "manly form" and comely face,
He bore with manhood's budding grace;
And his voice, tho' low, was deep and full,
As he sang with the maid on the milking stool.

He carried the pail, o'er-topped with foam,
To Angeline's vine-covered bright "happy home,"
Then through that soft and star-lit sabbath eve

   He sat on the porch by young Angeline's side;
Bright visions of hope together they weave,
   Ere the "launch on life's dark fathomless tide."
They part when the waning moon rose bright,
And shed o'er earth its "silvery light"—
If you wish to know what vows they made,
What deeds they did, and what says they said,
You must ask of somebody wise in such lore,
For I wasn't there, and can tell no more,
Save that ever when seven days wheeled their flight,
And bro't again the Sabbath night,
Did Oscar sit by that cottage fire,
And list to a tale from the grey-haired sire;
And tho' long and tedious the tale became,
To Oscar's ears it was still the same;
For no heed could he give to the tale of strife,
He could only think of his future wife.
Tho' mute and attentive he seemed to be,
His thoughts were working busily,
As he built a castle of azure air,
In which to dwell with his chosen fair.
Thus time flew by, as he always will,
Let us wish e'er so much to make him stand still,
And Oscar so long had worked in the mill,

That his earnings would buy him a snug little farm;
So away to his lot in the woods went he
To fell and to burn each tall forest tree,
    And make him a cabin tight and warm.
Thought he, "when' tis finished, and snug, and neat,
I'll throw myself at Angeline's feet,
    And tell her to name the day;
Then hither we'll hie from the world away,
And spend in labor the hours gay,
    Till eve brings a season of quiet rest;
    And then together in our nice little nest,
We'll sing some ballad or 'olden lay,'
    And be 'purely and perfectly blest.'"

PART II.

The Afterclap ignored by romances—being the main body
of life's matter-of-fact.

'Twas done—the hymenial knot was tied—
The maid of sixteen was a bride;
The groom was of a riper age,
For twenty years had made him sage;
That *sagesse* made him build a mill
Upon his winding mountain rill;
Hard by be reared his humble cot,
Nor envies kings their higher lot.
Thither his slow ox-team he drives,
    Dragging behind a groaning cart,
Which shrieked as if a cat's nine lives
    Were in it, struggling to depart.
No wonder; cart of mortal make
Could not endure it—it must break.
Just as it jolted o'er a stone
    High swelling in the road,
The cart gave one prodigious groan,
    And fell splash in the mud!
The axle broke, what could he do,
Unless he sunk in the mud-hole too?

Despair seized on him for a time,
But loosed his hold as Angeline
Proposed the furniture to save
From sinking in a miry grave.
He seized the tho't—with all his might
He threw the goods both left and right,
Resolved to resurrect the cart,
   He made an axle-tree of wood,
   Which, if it wasn't very good,
Still served to cheer his heart,
And bear his replaced chattels on,
"Over root and mossy stone."

Tho' the honeymoon was over,
Oscar Rose was still a lover,
And felt a lover's full vexation
Because it rained "like all creation,"
And March winds blew their fiercest chime,
To welcome home his Angeline.
"Ah, love!" he cried, "what shall we do?
The roof leaks streams and puddles too,
But here's a place quite dry for you."
"Oh, well"; she said, "I'll go to work,
And cook potatoes, eggs, and pork,
   If you'll set up the stove—"
"Yes, yes!" and then to work he went,
   While rain his garments soak;
But part of the stove was badly bent,
   And part was fully broke;
But they fixed it up, and made it do,
And Angeline cooked a smoky stew,
(For the cracked stove leaked both smoke and ashes,
   And the wind in fitful gusts
The rain through every chink hole dashes,
   And blew in pecks of moss and dust.)
They ate it (not the *dust*, but stew),
And went to work with zeal anew.
Oscar o'erhead stopped out the sky,
The cracks were stopped—the floor mopped dry—
And then fatigued beyond all measure,
They went to sleep and snored with pleasure.

Next morn the sun rose clear and bright,
They eat their meal with appetite,
And Oscar hastened back to town,
    To fetch another load;
This time he didn't quite break down,
    Nor upset in the road.
So everything was "fixed" all right
Before there closed another night;
So they were settled in a trice
Snug as a nest of tiny mice.

Oscar's father died, and his stepmother came,
With her mother who was blind, and deaf, and lame,
    To live with him and his deary—
Tho' deaf, not dumb, was Grandame Hill,
For her tongue ran on, like the cork leg still,
    Tho' all ears with din were weary;
Then she was so cross she couldn't be endured,
Excepting because she couldn't be cured.
    The stepdame was one vast interrogation—
Minding all business that wasn't her own—
She would whine all the time to make her griefs known,
    And scold with a voice like the Bull of Bashan;
Then talk of her pious faith and labor,
And then, next minute, slander her neighbor.
Angeline had six "sweet little dears,"
That oft got boxed on their "sweet little" ears,
By the stepdame's grandmotherly hand,
Because in the way they happened to stand,
Or could not reply to some interrogation
That she made in her tours of investigation.

No secret cell did the house afford
Where Angeline could hide a secret hoard,
    Or where aught could secret be;
For every drawer, and trunk, and chest,
Was an object of curious interest,
    The stepdame all must see;
If opposed she'd sigh as if wounded to death,
And say with a kind of sobbing breath,
"Ah, why am I doomed on earth to live,

And no respect from my child receive?"
Then she would cry and sulk for a week,
And her malice dire on the children wreak.
Then she'd a daughter, a squint-eyed girl,
   Who'd married, in spite of her friends,
An ugly, piratical looking churl,
Who, to make most ample amends
For all his numerous peccadillos,
   And to appease a fit of rage,
   Took an anonymous pilgrimage
Across the foaming Erie's billows,
And ne'er returned to Sylvia more;
But left to heal her widowed heart,
That his desertion might make sore,
   A small edition of himself,
   A little, silly, roaring elf,
That Sylvia's switch oft caused to smart.
Her husband's debts his goods consume,
So Sylvia had no house or home;
So trusting unto Providence,
And her step-brother's benevolence,
She took herself and darling boy,
His bounteous mansion to enjoy.

If Sylvia's baby mightn't run over
   Every child of Angeline's,
She'd tell Dame Hill, and her own cross mother,
   And then would follow a chorus of whines.
If Oscar and Angeline ever went out,
   Mrs. Rose and Sylvia'd pursue them—
"Why, children! what are you talking about?"
Mrs. Rose would say, and begin to pout,
   "I wish the world only knew them!
They treat me as if I were only a stranger!
   Stop talking if I but come nigh!
I believe they'd grudge me hay in a manger!"
   And then she'd begin to cry.

Dame Hill would tell monstrous long stories,
   So loud that herself could hear,
About her good man's fights with the tories,

Her husband, that "poor lost dear."
She had placed goods in Oscar's possession,
   Worth some thirty dollars and odd,
And she deemed it a horrid aggression
   If Oscar hung not on her nod.
"Ingrate! I brought him his wealth,
   Now he won't mind a word that I say!
Oh! if I could but have my health,
   I'd go with my money away!"

"And just see how poor Silvy is used!"
   Mrs. Rose then put in her word—
"I never saw girl so abused!
   Of such abuse I never have heard!

She helps Angy wash, and takes care of her child
(Just see what a heap on the table they've piled);
She takes care of the chambers, and waits upon you,
And all this she does without wages, too!
Poor girl! because she's unfortunate,
Such advantage as this of her misery take!
Oh! shame on the wretches who thus can doom
An orphan widow to poverty's gloom!
Make use of her labor, and when 'tis done,
Give her only support for herself and son!"

"Yes, shameless indeed!" cried Grandame Hill,
"Oscar has a farm, and house, and mill,
And money can raise when he wants, if he will—
   Why don't he reward her well?"
Dame Hill had heard what her daughter said,
For she spoke in a tone that might wake the dead,
   And shame St. Paul's church bell—
Then both the old "ladies" began to cry,
And wish that they might quickly die—
   Then scolded, sighed, and sobbed all at once—
Dame Hill must hear every word that was said,
Or else she'd take it in her obstinate head
   That they thought her a deaf old dunce;
So she was apprised of all that went on,
Tho' her sight, and hearing, and strength were gone!

There was a bachelor uncle—a drunken sinner—
Was too stingy to afford his own bread and dinner,
Half the time loafed at Oscar's good table—
Gave orders about managing farm and stable—
Taught the children to smoke, and drink, and chew,
And say with each breath the innocents drew
    Some profane or unseemly word;
Until the parents turned him out,
And he left with an injured air and pout;
    And the next thing that they heard,
Was that the wretch had suddenly died,
And left his wealth to Frederic Hyde,
    Sylvia's idiot baby!
'Twould be twenty years before he could touch
The hoards that cost his uncle so much,
    Yet Sylvia now thought her a lady!

Oscar and Angeline lived in fear,
Lest all their words Dame Rose should hear;
They dare not talk till she was asleep,
And then awake they couldn't keep,
'Till one Sabbath they strayed to a beechen grove,
To talk of the days of their early love.

"I had dreamed," said Oscar, "of a happy lot,
With friends surrounding our little cot,
    Ourselves doing good and no evil—
And we *do* do good—there are grandma and mother,
They scold enough to fright the devil,
And we bear it and let it pass over.
I thought when we married and moved here home,
That we would quietly live alone,
    Enjoying felicity pure;

'Stead of that, what quiet or peace do we get?
Grandma's in one perpetual fret,
And mother, who sees, torments us more yet.
    What evils we have to endure!"
"Then mother scolds," began Angeline,
"If I dare to think that the house is mine—
If I don't do as she bids she begins to cry,

And say, 'Of course you know better than I;
I'm but an old fool you all despise!"
And she pours out a torrent of sighs,
   And grumbles and scolds all the time.
Then she knocks round the children as if they were wood,
That she wanted to hurt if she possibly could,
And complains of Sylvia's hard lot—
I ventured to hint we'd build them a cot,
Where they in tranquil peace might rest,
Where the children would ne'er their quiet molest!"
"That's the thing!" cried Oscar, "The very plan!
I'll build the cottage as soon as I can,
   And be rid of this troublesome pest!"
"Stop! stop! not so fast!" cried Angeline;
"When I told her this she began to whine,
And vowed she knew full well before,
We wanted to drive them from the door!"

"Well then, let's endure it as long as we can,
For mortal life is but a span—
Sylvia may marry—the old woman die—
(I'm sure if they did I shouldn't cry)—
And then we might live in comfort and ease,
With our children around us in quiet and peace;
Were we rich we might furnish ourselves a room,
Secure from a witch that could ride a broom,
   And these step-dames' tongues defy!
But henceforth let none but a fool in his dotage,
Dream of comfort and peace with 'Love in a Cottage!'"

THE DUAL SPIRIT.

MEN lived in tents, and tended roaming herds,
    Ere, sprung alike from earth and heaven arose
Twin brother spirits, that with majestic words
    Beckoned a palace from the quarries close,
And, fashioned out of common clay, they chose
    A puny mortal, and nicknamed him king.
A waste of splendor round him they dispose;
    And other mortals came to worshipping
    That mortal for his show and glittering.

And yet through human hands alone they wrought,
    To dig out treasures from the darkling mine;
To sculpture monuments of pride and thought,
    Or frame the lyre, or pour the song divine;
Or do great deeds the numbing hand of time
    Would pass untouched when thrones had crumbled low;
And mingled with oblivion's wasting slime,
    They loved alike to raise or overthrow,
    To toil 'mid torrid sands or everlasting snow.

And they were loved and hated, praised and jeered;
    Oft victims both of flattery and scorn;
Dreaded and courted, idolized and feared,
    They dared the wrath of foe, and time, and storm.
Men blest and cursed the hour when they were born,
    And yet earth blossomed and the low wind sighed;
They thawed no snow and chilled no sunshine warm,
    And yet have men arisen, lived and died,
And scarcely knew when with the twain that there was aught beside.

Ambition! Genius! How your names can flash
    Electric fire across the sluggish soul;
As in the stagnant air loud thunders crash,
    And shake the welkin with their sudden roll.

So oft they burst thro' custom's strong control,
    And shake the spirit with their lofty tone;
And from equator to each distant pole
    Earth yields them tribute from each circling zone,
    And man has bowed, a willing slave before their glowing throne.

But once when men had caught their burning flame,
    And taught its light on loathsome scenes to glow,
Burned hamlet, city, palace, hut, and fane,
    And their mad shouts rose up as if to show
How men could joy in utter human woe.
    When priests had thrown their sacred garb around
Each hideous sin mankind can ever know,
    When sweet humanity's soft voice was drowned
    In the great war of warring hosts, and discord's jarring sound.

Then Truth in sorrow veiled her radiant head,
    And Falsehood reigned above a darkened world;
When pitying Mercy wept unnumbered dead,
    From out fierce tyrants' bloody pathway hurled;
When up from human fuel slowly curled
    The blackening smoke of persecution's rage,
The twain in sadness then their pinions furled,
    For Hist'ry 'd written on her bloody page,
    "Ambition lit War's wasting torch, and led his pilgrimage."

And men arraigned him at their judgment bar;
    The great blind multitude, who only knew
That 'mid the clamor of each feudal jar,
    Or war of nations, or of robber crew,
Ambition's torch o'er all its red light threw,
    And loud, 'mid clarion's, and 'mid trumpet's bray,
His voice of thunder rolled and echoed through;
    They knew not that the light was stolen away
    From hallowed altars, and profaned amid the fray.

They knew not that that pealing tone
    Was sounding ever on thro' space and time;
But unto glowing, high-wrought souls alone,
    Pealed forth its anthems or its triumph chime.

But those who caught it often linked with crime
    The words of power that they caught from Heaven;
They charged to Him, the generous and sublime,
    The evil that man's crime alone had given.
    They banished him, and forth that glowing soul was driven.

His brother stood unnoticed in the throng,
    His dark eye dimmed with sorrow's scowling cloud
He loved the right, but felt the power of wrong,
    Saw it, and heard it, 'mid the surgings loud
Of that great monster, the wrong-headed crowd;
    Yet not by anger, nor by pride, nor fear,
Was Genius' spirit for a moment bowed;
    Half-wept he one disdainful, pitying tear,
    Then with his brother took his flight half round the sphere.

A darkness settled on the blind old world,
    A mist of apathy and stagnant gloom
From dull, inactive slothfulness, upcurled,
    And mould and mildew grew on glory's tomb.
And science' buds were blighted ere their bloom,
    And art was banished 'mid forgotten things,
To mourn the glory of her by-gone noon;
    And music snapt her sweet vibrating strings,
    And poesy 'mid deepest shade furled up her shining wings.

But fast and far across the waters blue—
    The rolling billows of the ocean's surge—
The twain in silence yet unwearied flew,
    'Till they had touched its forest-tasselled verge.
No more their flight so constantly they urge,
    But paused, and glancing where they late passed o'er,
Saw from the flood a pall-like mist emerge,
    And float off darkly towards that distant shore
    Whence they had come. They sighed, and turned to flight once
        more.

Fate drove the exile far across the land,
    Deep to the bosom of a giant rock,
And chained him down with stern despotic hand,
    Where his cold dungeon evermore would mock

His restless spirit, while the tempest's shock
　　And zephyr's murmur far above his cell
Made not a sound to pierce the solid block,
　　Where strong Ambition had been doomed to dwell,
　　Where Genius followed him, self-exiled when he fell.

And they were left alone, and darkness stood
　　Stiff, palpable, and visible around,
Above—beneath—while all her monster brood
　　Of hideous fancies throng the gloom profound.
How fared he then—that exile doomed and bound?
　　The proud, aspiring, progress-spirit then?
Hope could not breathe one single cheering sound,
　　And chill Despair came o'er his life again,
　　Though oft repulsed in scorn—like all his faith in men.

Ne'er on his path had Hope refused to smile;
　　Ne'er had he faltered or despaired before;
Thwarted and crushed, had lain in dust awhile;
　　Then like a whirlwind on a tropic shore,
Burst he to life and energy once more.
　　In life's rough discipline, in school or field,
One word he ne'er had conned or pondered o'er;
　　Against one lesson had his heart been steeled,
　　He could not, would not learn the bitter words—*to yield*.

But now an hour of dull and death-like dread
　　Came o'er his soul—a with'ring agony,
Such as men feel who have no tears to shed,
　　When they have drunk such drops of misery
As few can live to bear—he breathed no sigh;
　　He spoke no word; not e'en a thought broke thro'
That stunning dread. He knew he could not die,
　　And yet no hope for future years he drew.
　　He had been conquered—this was all he felt or knew.

A voice whose silvery sweetness softly fell,
　　Like the low ripples on a shell-strewn shore,
Or dying echo of a distant bell,
　　Broke through his trance—his gush of passion o'er,

He started up—a spell those accents bore
　　To fright despair. "Look up!" those words were all
That made Ambition like himself once more—
　　He stood within a lofty lighted hall,
　　While round him rose a thousand pillars white and tall.

Bewildered—glad, and awed almost to fear,
　　He stood one moment; bursting then aloft,
Rose his wild shout that echo loved to hear,
　　Catching it up and uttering full oft,
The joy cry loud, till whispering low and soft
　　The murmurs far in distant music die,
"I am not conquered! tho' exiled and scoffed
　　By man's decree and blind old destiny,
　　'Twas but a moment that I stooped enough to sigh."

He paused till echo rolled and died away,
　　Like an imprisoned thunder sunk to sleep,
And stillness followed its tumultuous play,
　　Heavy—profound—like death's abysmal deep,
Whose gates the voiceless shades for ever keep;
　　Awe fell upon his spirit and he spoke,
Subdued and soft as summer zephyr's sweep—
　　"Was that thy voice, my brother?" yet it broke
The spell of silence, and a thousand whispers woke.

"My brother!" "brother!" "thine my brother!" "thine!"
　　Mingled in murmurs, wildering to hear,
Confused and strange as sounds the morning chime
　　Of wild birds warbling their heart-stirring cheer
A flood of clashing music full and clear,
　　Upon the wondering and repeating air,
Till the vibrations half beguile the ear,
　　To think no music so surpassing rare,
　　As those wild echo-glees that woodsmen only share.

But mingling with those waves of echo came,
　　A voice he knew could not be all his own,
Familiar accents spoke the exile's name,
　　With deep affection in its low-breathed tone:

He knew—he felt—he was not all alone—
    The proud exultings of his heart grew still,
A half-shed tear within his eyelids shone—
    A world was powerless to break his will,
    And yet rone word could thus his stubborn spirit thrill.

"Now who is thwarted, Destiny or thee?
    Where is thy cell, of earth thy destined part?
Can bungling man inspired by golden fee,
    Carve such a palace from the mountain's heart?
Can their ideals thus to being start?
    Their glowing thoughts congeal to solid stone,
Without the tardy aid of feeble art?
    A city might lie dwarfed beneath this dome,
    That by my will I built for love of thee alone!

"O brother! with a heart of earnest love,
    And strong free faith in truth's unfailing power,
With will unflinching as the stars above,
    One well may balk earth's most exultant hour,
Of half its triumph. What tho' Fate may lower,
    And men obedient echo back her ban?
The wrath of mortals, like a summer shower,
    Falls and is past; and hate's most bitter plan,
    Dies with the failing life of poor, short-sighted man.

"Thou art a strong, and I a wayward one,
    Without thee, oft I ramble bootless where
Thy eagle eye and spirit-stirring tone,
    Would rouse up hosts thy energy to share;
Thou from the desert, and the mountains bare,
    Would force out action by incessant toil,
While I, too oft in valleys rich and fair,
    Riot mid nature's most luxuriant spoil,
    And leave unwrought the strength of that unbroken soil.

"But when with thee, for me the deserts bloom,
    I make a palace of a dungeon cell,
Despair's dark night with Hope's wild fires illume,
    And change doom's mutterings to the wavy swell

Of music watching as a fairy spell;
    With me thy footsteps blaze with glowing light,
And like the chime of a cathedral bell,
    Thy voice, o'er ocean surge and mountain height,
    Peals forth in thunder tones thy words of might.

"Henceforth let us be one, a Dual One,
    That, whom thou lovest, him I too shall love—
We'll reap together what we both have strewn,
    And strew together wheresoe'er we rove;
But should I tarry in some myrtle grove,
    While far beyond thy prouder footsteps stray,
Then, whom I bless thy spirit shall not move,
    But he shall be like those red lights that play
    O'er northern nights—as aimless and as vague as they.

"Bird-like he'll tune his nature-prompted song,
    Nor seek to leave his native tangled shade,
No motive power shall urge his course along,
    By impulse started and by impulse stayed,
Prompted by feelings that he never weighed,
    One hour he'll sigh o'er some imagined ill,
And joys, like Autumn leaves, will dim and fade,
    And then as sudden, with a joyous thrill,
    Pleasure shall crown life's cup that all the graces fill.

"Unknown to me some mortal shouldst thou name,
    That soul shall feel Ambition's force alone—
Will toil from land to land—from main to main,
    To carve his name on rocks of every zone;
Each desert rock to him shall seem a throne,
    Seen thro' hope's mirage, ever drawing near,
His barren path shall ne'er be overgrown,
    With bloom I strew as I have strewn it here!
    He shall be hard and cold and held in awe and fear.

"But he to whom we both our blessings give,
    Shall mow the forests like the blooming grass,
Or plough the mountains or subdue the wave,
    Or far aloft the wilds of Æther pass,

Or touch the pulses of the human mass,
    And count its heart throes and find out the word
That rules like magic o'er the populace vast—
    The word of power stronger than the sword,
    And where he listeth he shall rule the people as their lord!"

How the Dark Ages slowly rolled away,
    And Destiny recalled the exiled pair,
How new life struggled with the old decay,
    And new hope wrestled with the old despair,
Is it not writ in Hist'ry's pages fair?
    And they who led the exodus from night,
Who brought up buried truths to upper air,
    For blindfold science, who found out the light,
    Were they not gifted of the twain—the Strong soul and the
        Bright?

A WORD TO THE WEARY.

THE eyes of men are failing with looking up so long,
For fleetness to outrace the swift, and strength to fight the strong;
And their faith is full of doubting, when God's lightnings seem in vain,
To melt the ice from human hearts, or rive the bondman's chain.
And the lip hath learned to murmur at the law ill understood,
Which sends the sun and rain alike on the evil and the good;
And the heart grows sick and weary that the seed so freely sown,
So often falls on barren sand, and cold unyielding stone;
That words return so dull and void, that went forth strong and full
Of truth's most earnest prophecies against oppression's rule,
And when he, like the seer of old, would call down scath and fire,
Heaven's blue serene but seems to mock his soul's impatient ire.

He sees a nation's garments grown, by threat and lash-paid toil;
He hears a murdered brother's blood cry ceaseless from the soil;

He sees no forest deep enough in all our wide-spread land,
To shield the crimeless fugitive from the oppressor's hand.
A stranger rests within thy gates—a weary one is he—
What boots his name, or hue, or race? What matters it to thee?
Ahungered, cold, athirst, he comes—thy bounty gushes free;
A Saviour's tones approving say, "Thou doest it to me."
But hark! is that a bloodhound's bay? Why comes its howling here?
Why turns the stranger's tranquil brow aghast with sudden fear?
O! woe for him, that hunted one! thou canst not give him rest;
Thy household altar cannot shield that "welcomed stranger guest;"
The Marshal and his minions full armed are on his track,
To his loathsome house of bondage to bear God's image back.
No homestead walls, no fireside hearths, are sacred from the hands
Of human vampyres such as these, who scourge our slave-curst land.

What wonder that men's souls grow sick at all this woe and wrong?
What wonder that their faith is weak, when Evil seems so strong?
She boasts her strength with voice of pride, and down before her fall

The leaders they had watched and hoped would break her iron thrall.
Then trust not to such wandering stars, to *each* a soul is given;
Each has his own peculiar work to do for earth and heaven.
We read a fallen leader drew a third of Heaven's bright host
Down to that grim and black abyss, where all were merged and lost.
Then follow not blind, erring guides, when God can lead us all;
We owe allegiance to none else, we need no other's call.
The arm of flesh is faint and weak, and if the spirit fail,
O! when in all this outraged world will truth and peace prevail?
When the half-wrecked ship is drifting upon the stormy sea,
The sailor has no time to pause and wail despairingly.
Storm-worn and dripping with the brine of many a billow's crest,
The weary battler with the waves can find no time to rest.
And now this whelming tide of wrong the bark of truth has tossed,
Till hope's dim beacon, 'mid itswrath, seems well-nigh drowned and lost;
Yet all the more with earnest grasp hold fast the driving wheel,
And steer amid the breaking surf our good ship's stubborn keel.

The weary must not heed their toil, nor weak ones count their fears;
The mourner must forget his grief—there is no time for tears;
The night watch must not miss our work, or when this driving storm
Shall break before the blazing wheels that bring the coming morn,
We shall not see the riven clouds fall dead and prone apart,
And from their very blackest midst the genial sunlight start.
Take courage, weary toiler! Heaven whispers earnest cheer;
For ever when we least expect, the light and help draw near,
For we know that God is with us—that for the *man* he gave,
He will not take from master's hands a chattel soul—a slave.
Ah, spoilers! tremble when he calls, for that manhood robbed away;
For the host of rights and duties you are snatching from your prey;
Oh! soldiers, on truth's battle-field, with your arms of love and light!
I charge ye, faint not—God has pledged the victory to the right.

BY THE MISSISSIPPI.

LIKE to a lake the stream before me lieth,
　　Between a rocky and a wooded shore,
Cradling the many tints where daylight dieth,
　　Repeating each expiring glory o'er.

And there the forest shadows shake and shimmer,
　　In every ripple by the breezes curled,
All shapes are mingling in one common glimmer,
　　Within that mirror's strange inverted world.

A white sail, like a sea-bird o'er it glideth,
　　And floats behind the island's twilight grey
Where creeps a pale mist slowly up and hideth
　　Whatever lieth in its milky way.

The blue Itasca would not know her daughter,
　　So soiled with travel and so overgrown,
Who spreads yon rolling sea of turbid water,
　　A thousand leagues out toward the burning zone.

From out the caverned bluff the night bird calleth,
　　And from the quarry comes an answering cry,
The flitting double of the night hawk falleth
　　Down where the other shivering shadows lie.

And one white star on night's cool forehead glistens,
　　And glistens upward from the waves as well,
And with hushed breath my very spirit listens
To some sweet music's far-off, dying swell.

I almost wish that I could do my sleeping
　　With yonder wild birds on the river's breast,
And with the stars their ceaseless vigil keeping,
　　Be like an infant once more rocked to rest.

SACRAMENT.

Up by the old road on the hill,
 Now with the creeping grass o'ergrown,
Near where the faintly springing rill
 Is oozing o'er the mossy stone—
Hard by the grouse's deep retreat,
 Through all this soft autumnal haze,
Wait for me at the old time seat,
 Dear Love of by-gone days!

The world is flushed with ceaseless change,
 Its brimming joy cup running o'er,
But weary of the new and strange,
 I long for our old haunts once more.
I long to tread each dear old hill,
 Beneath its woods of red and gold,
And 'mid their riches grand and still,
 Be soul-baptized within the Old.

Cease shivering, trembling Aspen tree—
 I would not have thy rustling heard!
Murmur not now, thou Pine wood sea!
 Sigh not, soft Wind! chant not, sweet Bird!
Lay thy dear hand in mine, my Own,
 That I may feel thy heart throbs near,
And take from God, with me alone,
 This Sacrament of Silence here.

MAKE NOT POYERTY'S CUP TOO BITTER.

Ye may harness the lightning till trained like a steed,
It will carry your thoughts with its limitless speed,
Ye may yoke the fierce whirlwind till bowed to your will,
It will grind like the ox you have broke to your mill;
Ye may tame the wild cataract's flood till it feels,
Like a felon condemned at your factory wheels;
But there stay your strong hand, nor dare lengthen your chain,
E'er to harness a *soul* to your engines for gain.

Ye may desecrate Nature, and haughtily tread
On the wrecks of its beauty, disfigured and dead,
But 'twere better for you that ne'er from the sod
Ye had started to life, at the mandate of God,
If ye dare to subdue to your power the will
Of a soul, which, tho' crushed and distorted, is still
In the image of Him, who hath, equal and free,
Made that spirit, proud atoms of frailty, with ye!

We can toil for a purpose, and cling unto life,
Thro' its storms and its turmoil, temptations and strife,
While the purpose is high, and the motive is pure,
Few indeed are the trials we cannot endure.
But go—put your curb on another's free will,
Keep us back from our aim, yoke our souls in your mill,
Make our poverty something too bitter to bear,
Ye will see then how much a high purpose will dare.

We can sleep upon straw on the cold garret floor,
We can toil on the crusts Dives casts from the door,
We can shiver half clad by the unlighted hearth,
Tho' we quail at the sound of the northern wind's mirth.
We can bear it and smile if the heart food is there,

To urge on the free spirit to do and to dare;
Aye! can laugh at privation, and hunger, and cold,
And thro' scorn, tho' in rags, be strong-hearted and bold,
For we know that the wealth of Peru cannot buy
Our ambition's proud hope—or a home in the sky;
And each slight that we feel, and each sneer that we meet,
Adds fresh fire to our hearts, and gives wings to our feet;
Every force that opposed, when o'ercome makes us strong,
Gives us courage to battle with famine and wrong,
But dare bar up the path to our purpose, and see
If the poor in their might are not stronger than ye!

TO-DAY.

A GOLDEN cloud with silvery traces,
    The *future*, floated on before,
Its border smiled with angel faces,
    A double rainbow spanned it o'er.

The sunlight thro' its winged edges,
    Shed a dreamy, mystic calm,
As o'er the mountain's serried ridges,
    Unscarred by peak or crag it swam.

O'er my childhood trailed its shadow,
    As unchildlike sad, I lay,
Beneath the old tree in the meadow,
    Listening to the water's play.

And time's wooing winds were winning,
    All my long days thitherward,
Unthought-of stretched life's path of sinning,
    Pain-enhedged and fire-scarred—

The future gathered dark before me,
    One grim cloud, streaked with sulph'rous dun;
It came to meet, it shadowed o'er me,
    And life's battle seemed begun!

"O my dreaming!" cried my spirit,
    "Is human life a storm like this?
Must I from childhood's gloom inherit,
    Maturer grief for hoped-for bliss?"

"O! thou golden-clouded future!
    Angel-winged and rainbow-spanned!
Didst thou herald woe and torture,
    A mocking mirage 'mid life's sand?"

But that scathing future came not,
    Where fell its shadow, fierce and dread,
And my youth but shared the same lot,
    That my weary childhood had.

And my spirit felt such yearning,
    For a wider path to tread,
Its long task of patience learning,
    While it knew not where it led.

And then the future, like a morning
    Blank with fog, before me lay,
'Till my searching soul took warning,
    And turned its scanning to to-day.

Lo! my cramped path expanded!
    Roses flushed its hedge of thorns!
The free air my soul demanded,
    Blew fresh with breath of summer morns.

Then my soul looked round in wonder;
    "Whence," it cried, "this brightened way?
Was this the promise of the thunder
    And the clouds of yesterday?"

"Have my cheerless footsteps led me,
    Footsore up the mount of toil,
Where no friendly aiding sped me,
    To tread, at last, enchanted soil?"

The cloudlet of my childhood's vision,
    Around me spreads its hazy gold,
While onward, seeking realms elysian,
    The angel hosts their wings unfold.

How their great white pinions beckon,
    As on o'er gulf and crag they go!
Bridging chasms, thunder-stricken,
    With my child-dream's promise bow.

Then farewell, Hope! and farewell, fearing!
   Farewell, sighs for yesterday!
Since angel guards of faith and cheering
   Make life rich enough to-day!

SUMMER FRIENDSHIP.

"I LOVE thee!" the Apostle said,
Yet, ere another sun had shed
Its flood of light on tower and fane,
That friend denied the loved one's name,
And one more false and treacherous sold,
That priceless life for counted gold;
Yet, who more loving seemed to be,
Ere that dark hour of wrath, than he?

And yet the one they crucified—
The one his loving friend denied—
Was kind and gentle and more true
Than other friend earth ever knew,
Yet he found friends whose love grew cold,
Friends who denied him—friends who sold—
Friends who were never friends indeed,
And left him in his utmost need.

If false ones smote that gentle one,
So kind to all and harsh to none,
What erring child of earth may claim
A friendship deeper than a name?
Tho' all may *seem* to love thee well,
Who, for the coming hour, can tell
Which of the friends who love in this
Will not betray him with a kiss?
Who, ere life's changing sands are run,
'Mid all his friends is sure of one?
Who, ere life's journey hath an end,
Will dare to say, "I have a friend?"

BY THE SEA-SIDE.

I FELT the beating pulses of the cool and trembling sea,
      Thrill 'neath the silent touching of my earnest hand,
And the long bright ripples ran in meeting me,
      And leaped in laughter on the shell-strewn sand,
And my soul gushed up and over, as at clasp of friendly hand.

The sea-weeds rooted close, and clinging to some old and wave-worn shell,
      Cast their green and dripping tresses underneath my careless feet,
And the fairy foam-wreaths mounting every long and wavy swell,
      Tossed their white plumes on the air, and rushing glad and fleet
Dissolved in rainbows on the waves that murmured at my feet.

And the old and mythic dreamings of strange mystery and might
      That whispered their enchanting spells to my early childhood's ear,
Came flashing o'er my spirit's gloom, like broken gleams of light,
      Leaping from cloud-thrones drifting wide in the far off upper deep,
And then as sudden fading back to darkness and to sleep.

Then the soul within me, harness worked and labor tried,
      Arose and shook its shackles from each bondage wearied limb,
New strength seemed waking in me, new fervor and new pride,
      The new glory dawning on me made all earth-born glory dim,
And life transfigured stood before me, true of heart and free of limb!

Then life's battle seemed a conflict, not of pain, and toil, and woe,
      But more a war of wind and wave, of tempest and of sea!
And a stormy love of combat, in my veins began to glow,
      And heave my heart in great pulsations, in the joy of being free!
O! my soul grew brave and stronger from the teachings of the sea!

I'LL TELL YOU, COZ!

You ask me if I ever loved, and I most freely tell,
I *have* loved long and faithfully, loved earnestly and well,
Loved all things good and beautiful with love like worship given,
Because I could not lock within the richest boon of Heaven:
Bind up the wind in some dark cave, and chain the restless sea,
Then prison in the human heart its springs of sympathy!
How can you ask of one like me, who on the earth so long
Has journeyed on life's pilgrimage, with all its changing throng?
O ! I have loved, and there are none in this dark world of care,
Who in my heart's great sympathies have never held a share.
I've loved the tender and the cold, the worthless and the true,
Within my heart a bud of love for each thing human grew,
All things that breathe upon the earth, and e'en the very air,
The guiltless and the beautiful I have loved everywhere,
And yet in all this realm of love my heart has been a throne,
And he that sat thereon has ruled my spirit-land alone,
And obedient to my heart's enthroned, as to a god's behest,
I've loved the world, the whole wide world, but my own *will* the best.

PITY.

How much is pride, that humble seems!
  How much is false, truth's guise that wears,
How patronizing kindness leans
  Its crushing weight on him who bears
The hateful mockery, as the saint
Bears fire and stake, without complaint.

'Tis hard in gratitude to bow
  And teach the stubborn lip to lie,
When the heart's writhing pride must glow,
  Insulted, in the cheek and eye,
Yet utter *thanks*, when it has borne
Man's stupid pity—worse than scorn!

It's not enough the struggling soul,
  Debarred from paths it fain would tread,
Must bow to fortune's stern control,
  And barter life for bitter bread—
Bitter with scorn's cold wormwood dew,
But must it bear man's pity too?

TWAIN

How strangely things are mixed and blended,
　　Within the chaos of the brain!
Things unbegun with things long ended—
　　Some of joy and some of pain.

Things as unlike as things well may be,
　　Together through the soul's obscure,
Half float and sway in drifts unsteady,
　　Like snow-wreaths on a wind-swept moor.

Just now a friend I made but newly,
　　An old man, meek and quiet, came
Across my thoughts, which, all unruly,
　　Have called him by another's name.

And oft of late this name he beareth,
　　Within my spirit's unknown land,
Though ever his own form he weareth:
　　What link between them made this band?

My first friend's eye was like the eagle,
　　His frame with youth's strong sinews strung,
His voice, like call of silvery bugle,
　　Around its strong-willed message flung.

A fireside look, and voice, and tread,
　　My new friend hath, so sweetly human,
I turn and look at his snowy beard,
　　And wonder he is not a woman!

Why should that name come thus, unbidden,
　　To call on one so different far?
Unlike as caves from sunshine hidden,
　　And golden-blossomed prairies are.

They never met nor uttered greeting,
    Never heard each other's name—
Why in my thoughts do they keep meeting,
    As though they somehow were the same?

VOICES.

Like the ripple of bright wavelets,
    On a beach of soft sea-sand;
Like the flutter of the leaflets,
    In a wide-spread forest land;

Like the footsteps of the rain-drops,
    In their dance upon the roof;
Or the ringing elfin echo,
    Of a fairy charger's hoof;

Like the hushed hive's low humming,
    When night's solemn rest has come,
Are the whisperings of our spirits,
    When we seek to be alone.

We listen to their murmur,
    Till their pleasant rustling seems,
As full-voiced, deep, and real,
    As the songs we hear in dreams.

The unreal's changing shadows
    With the real's substance blend,
Till forgotten lies life's riddle,
    And its fast-approaching end.

For the mists of death's dark valley,
    Rainbow arches bright o'erspan,
And beyond, in boundless sunshine,
    We behold the path of man!

Then what matter for the darkness,
    Closed within our prison bars,
When we know, that past its portals,
    All broad space is bright with stars!

When we know a whole world's shadow
 On the wilds of æther cast,
Down before heaven's watch-light dwindles
 To viewless point at last!

And at last when life emerges,
 With the faint breath from our lips,
And its glimmering light is hidden
 By grim death's cold, dread eclipse—

Shall that darkness break the fiat,
 By the universe obeyed,
And thus be, 'mid life, the only
 Eternity of shade?

"Nay!" these thousand voices answer,
 From the earth and upper air;
From the inmost restless spirit,
 Outspeaking everywhere.

Death is but the gateway, closing
 'Tween the old life and the new;
Have ye never seen it open,
 When a passing soul went through?

Though sometimes the darkness lieth
 Dim beyond the outer wall,
And no star-beam's brightness pierceth
 Through that heavy midnight pall—

And we sigh for the soul unlighted,
 Going down that lonely way;
For our eyes are unanointed,
 To behold the future day;

Yet sometimes the sunlight streameth,
 Over all that solemn road,
'Till its glory casts a halo
 On the soul's cast-off abode.

And, oh! then these spirit whispers
   That have borrowed earthly tones,
Swell to such floods of music,
   As high heaven only owns!

BEAUTIFUL LIFE.

Swaying and swimming through sunset air,
Insect wings shivering everywhere,
Shake out a song from these quivering wings,
Which a dreamy sense of their blessedness brings;
How they joy in a being scarce measured by hours,
Since they die when the dewdrops come down on the flowers.
                    Beautiful life!

Down in the quiet pools sheltered and deep,
In their silvery mail do the fishes sleep;
Or up through the ripples they flash in the light,
Like a meteor's flame in a cloudless night;
Soulless, and sinless, and fearless are they,
While they revel in life as it passes away.
                    Beautiful life!

Cloudward and skyward upspringing, the birds
Pour over earth their musical words;
They rain down their music from out of the sky,
And from every thicket and brookside sigh;
As free as the songs of the waves of the sea,
Out pour their carols of rapturous glee!
                    Beautiful life!

Out from each shadowy cavernous glen,
Out from each meadow and marsh and fen,
Out from the prairie and down from the hill,
Come cries of the life blest, low toned and shrill;
From everything living, and everywhere,
Gushes life's joyfulness filling the air!
                    Beautiful life!

And human life silvered with morning light,
Golden with noon, or flushing toward night;
Grand with its conflicts of tempest and storm,
In its every hour and every form;
Mellowed and deep in its growing old,
How can its riches be fully told?
                    Beautiful life!

PHANTOM BUILDING.

A FOOTSTEP in the dust we trace,
    And then, of him whose step was there,
We build above that lowly place,
    A phantom figure in the air.

Lone Crusoe saw a shadow host,
    Hold savage orgies on the strand,
Because upon that barren coast,
    A human footstep pressed the sand.

The Arab bites his wordless lip,
    To see an armèd train pass by,
When nought of barb or "desert ship,"
    Save footprints, meets his searching eye.

We see a dead stalk on a wall,
    And suddenly to golden bloom,
There bursts through all its death-spell's thrall,
    The wallflower's phantom o'er its tomb,

A snowless winter's walk we take,
    Through some deserted graveyard old,
Where 'neath our feet the scentless brake,
    And grass, lie withered, brown and cold.

Their rustling crush recals their past,
    Like magic life-word to them spoke,
Their brown arms up toward Heaven they cast,
    Their wintry doom of thraldom broke.

Up from the dull and frozen mould,
    Transfigured springs the fragrant fern,
And verdant grass, and daisies bold,
    Smile round each solemn gravestone's urn.

We pass—some trailing brambles clasp
    Fast to our skirts with hookèd thorn;
We stoop to loose this tightening grasp
    Of stems, of life and verdure shorn.

We cast the rough encumbrance down,
    When full-leaved up before us rise,
With berries bowed these briers brown,
    Grown green and strong before our eyes!

Can summer skies melt bonds of death,
    With surer skill than this we share?
Dare magic words, in whispered breath,
    Evoke more phantoms than we dare?

What matter for the driving storms,
    The drifting snows, life's wintrier parts,
When in *us* live all glowing forms,
    Creative summer in our hearts?

"LIBERTY—EQUALITY—
BROTHERHOOD"

WHEN morning broke bright o'er the darkness that shrouded
The sleep-shackled world in a mantle of night,
When the wild wind had chased the dun vapor that clouded
The blue deep of æther obscuring its light,
When no prison confined, and no walls were around thee,
Hast thou never then felt that thou wast not yet free,
That the bondage of custom so closely had bound thee,
That other men's thoughts are a prison to thee?

Hast thou never grown sick in thy spirit with fearing
The laugh of some worm crept up higher than thou?
Or some pitiful fool, who had borrowed his sneering,
To whose broadcloth and gold all the multitude bow?
Yet thou dared not turn back, and with pride answer pride,
And with scorn spurn the scorn that fell heavy on thee?
Thy locks have been shorn, and thy hands have been tied,
False customs have bound thee! thou canst not be free!

Burst forth from thy bondage, proud spirit, and utter
Those truths that shall set other souls in a glow,
Let the lightning-bolt strike ere the deep thunders mutter,
Be free first thyself—then help all to be so!

Is not thy soul human, and life everlasting
A heritage free—by thy birthright thy own?
Who heirs an estate thine in grandeur surpassing,
Or who has a less, that thou ever hast known?

The laborer crushed in the dust shall awaken
From the sleep where he lost his dread portion of toil;
And the king shall leave earth when death's sceptre shall beckon,
Then the slave shall be peer of the lord of the soil!

No more shall the lordling in luxury silken,
Use the labor in which he has taken no part,
No more shall grieved echo sob back from the welkin
The sigh of the starved in mind, body, and heart.

For the grave owns but equals in dust it is 'tombing,
And souls find but equals beyond the dim shore,
Whose dark shrouding shadows, forbidding, lie glooming
Around death's deep waters we all must pass o'er.
Must the soul then, forgetting its own mighty powers,
Supine in the dust, put its birthright away,
Till the grim King of Terrors shall roll round the hours
To close up life's wintry and wearisome day?

Or shall it stretch forth its free wings, ere the dawning
Of life in the realms of the beautiful sky,
To meet the slow-coming, yet glorious morning,
Whose steps are approaching, whose advent is nigh?
When hand clasped in hand shall acknowledge its brother,
When peasant and prince shall be titles unknown,
When none shall claim rights he denies to another,
And all bow to one Monarch—one Father—one throne.

THE PICTURE ON THE WALL.

THERE'S a soft dreamy landscape that hangs on the wall,
    Through the sweet Sabbath solitude whispering things,
That a single light footstep in parlor or hall,
    Would drown with the echoes each footstep brings.

Some artist unknown with his soul all aglow,
    With a light other eyes had no power to see,
And a heart whose deep poetry could not o'erflow,
    For words were denied and his spirit not free;

Has here slowly evolved thro' his beautiful art,
    A shadowy scene from the dreamland within,
Where clouds shade that glory that poured on his heart,
    And his toil has brought twilight to soften its beam.

But thro' that soft light that glows dim on the lake,
    With its tinge on the foam and the archway grey,
On that brown ancient tower and tufted brake,
    And those clouds' ragged masses far floating away,

I can trace the pale glory of that spirit light,
    That but glowed for the soul of the poet alone,
And thro' those still shadows, foreboding the night,
    Are his fetters of soul to my spirit made known.

This glory and shadow so wedded in one,
    Give birth to a host of whispering dreams,
Who make silence their voice when they find us alone,
    To reveal the deep source of each life's hidden streams.

The landscape is lovely and brings to our view,
    The castles and rocks of some far away place,
But it shows us more faithfully, life-like, and true,
    The artist who painted those scenes which he traced.

ONE APRIL EVE.

I've been out in the "grand old woods" to-day,
  Where the earlier plant stems the damp earth part,
As their long chilled pulses begin to play,
  And their leaves toward the genial sunlight start;
Spring's first birds chirped on each budding tree,
  And merrily swung on each wind-swept bough,
And I never was younger, it seems to me,
  Or more of a child, than I am just now!

The frogs that were piping so shrill in the flood,
  Told the stories so oft in my hearing erst told,
And strangers were with me, the kind and the good,
  Who sang me the songs I had loved of old;
These spells have been breaking the chain-links of years,
  And sweeping me back to life's by-gone day,
Till my soul has swelled with its old time fears,
  Its loves and its joys that have passed away.

The fountains deep hid in my heart have gushed o'er,
  In tears of warm tenderness, spite of my will,
And I long for some dear one, familiar of yore,
  To catch its outpourings—its throbbings to still.
The kindness of strangers is touching and sweet,
  And gentle new friends for my gratitude call,
But I'd give all the world, this bright eve, but to meet,
  Some old friend I love—'twould be worth more than all!

BY A LAKE SIDE.

NIGHT's dusky plumes float slowly over,
    From the cold grey eastern sky,
Toward where eve's bright pinions hover,
    Lingering ere the day shall die,
And leave the chambers of the west,
    To this still, solemn, starlit guest.

Flaming chariot wheels of glory,
    Such as bore the seer away,
Blazed beyond the lake before me,
    As they carried off the day;
Then as the fading splendor fled,
    Night's dreamy spells were round me shed.

Across the lake and land and river,
    Those witching spells my vision bear,
Till with joy's quick, sudden shiver,
    Meet I one remembered there:
Yon blue horizon lost the key,
That locked the absent one from me!

O! bless thee, Night! for power given,
    To melt such weary leagues away,
And keep, where'er life's bark be driven,
    Oblivion's ghostly shade at bay,
Till those once near us seem so yet,
And we forget we can forget!

THE DEATH WATCH.

Night had spread o'er earth's surface,
    A damp and darksome pall,
Ocean sent up its vapors,
    To overshadow all;
Silence had hushed sound's voices,
    All save the night-bird's call;
The dull monotonous murmur,
    Of wind and waterfall,
The cricket's chirp, and death-watch ticking,
    Ticking in the hollow wall.

On the couch sleep had deserted,
    Tossed the restless to and fro,
Listened to the lonely night-wind,
    Wished for morning's rosy glow.
Wearied of the wind's low murmur,
    And the stream's unceasing flow,
Tried to marshal into order,
    Thoughts that from confusion grow;
Turned and sighed for gifts that mortals,
    On earth are destined not to know,
Then held the breath and stilled heart-beating,
    To hear the death-watch ticking slow;
Ticking in the wall at midnight,
    Monotonous and low,
A keeping time to death's sure footsteps,
    'Mid summer's bloom and winter's snow.

Then slowly creeps a shivering shudder,
    Through each limb and nerve and bone,
A thrill of half delicious terror,
    Exquisitely lone,

That amid the day's broad beaming
 We would blush to own,
But now with darkness round us brooding,
 Fear's dark power is known,
Superstition grasps our heart-strings,
 And makes the mind its throne;
And we think of hoarse death rattles,
 And the dying groan,
When we hear the death-watch ticking,
 Ticking in a gentle tone,
In the hollow wall at midnight,
 When we are alone.

OUR OWN OLD WOODS.

AFAR across the billows' sounding roar—
    The sky-like waters, and their star-like isles—
Why seeks the trav'ler on Italia's shore,
    'Mid clang of war, or old Rome's treach'rous wiles,
To find a land where genial nature smiles?
    Why search 'mid relics of forgotten art,
Where nature's tapestry drapes its ruined piles,
    For food to nourish in the mind and heart,
The latent feelings and the thoughts that may to being start?

Do Greece's statues to her noble sons,
    Who died in exile by decrees unjust,
Teach him to love his own land's honored ones
    Who lie unsculptured, mould'ring back to dust?
Do gods of stone teach him a holier trust
    In him whom hands have never shadowed forth?
Or crumbling thrones or monarch's trampled bust,
    Instruct his heart how little strength is worth,
More than the prostrate forest kings in lands that gave him birth?

Do nameless cities 'mid the desert lands—
    Once royal seats, but now the bats' abode—
Tell that lone pilgrim wandering o'er the sands,
    One tale untaught by any woodland road.
The schoolboy from his father's house once trod?
    When solemn awe his youthful soul o'ercame,
Where the green light thro' matted branches showed
    The narrow path,—his pent breath scarcely came,
Lest he should wake some monster wild, or shape without a name.

As in his mantle 'neath the open sky,
    He counts the islets in the upper deep,
Does not the jackal's ever wakeful cry,
    Or hoarse hyena, banishing his sleep,

124

Remind him of those days, when 'mid his sheep,
  His father's dog kept hungry wolves at bay,
When their long howls and many pattering feet
  Wore the dull hours of tedious night away,
'Till o'er that lonely wilderness the morn rose pale and grey?

The present danger melts and fades away,
  As mist wreaths vanish in the æther free;
For thro' the past his wakened mem'ries stray,
  His thoughts are busy far beyond the sea,
Where no red sands have buried bud and tree,
  And plain, and city 'neath their scorching waves,
Where wildbirds sing, where hums the honey bee;
  And many a wildfowl its soft plumage laves,
In glassy lake and river deep, hoarse rushing from its caves;

Or stream that gathers from each silvery rill,
  That winds through glen and narrow, dark ravine;
An added item its deep course to fill,
  While trackless wildwoods raise their trunks between,
And bathe their feet amid the waters sheen,
  And nod above the brink of each cascade,
Each leaning unto each its forehead green,
  To watch the motion of the flickering shade,
That their crossed arms, in airy dance, with the soft wind have made.

He fancies how that cabin, bare and rude,
  Might spread a shelter o'er his sleeping head,
Where beasts that prowl that houseless solitude,
  Awed by man's presence, would not dare to tread;
An Indian wigwam, or a hunter's shed,
  Were safe protection in that lonely place,
Where nations once in mutual struggle bled,
  Dreaming of glory, yet have left no trace
Of nation or of tribe—no name of chieftain or of race.

Here all alone, from feud and faction free,
  "The world forgetting—by the world forgot"—
The forest lord beneath some sheltering tree,
  Might search as deeply in the mines of thought
As he whose brain 'mid foreign scenes had wrought,

For gems of truth, where, 'mid his weary toil,
The very wild beast's desert howlings brought
    Back to his memory all the golden spoil
Of mental wealth he'd left unreaped upon his native soil.

By distance hallowed come the scenes of home—
    The rocky girdle of the ocean's surge—
The mountain's peak, the rushing cataract's foam,
    The tempest's shriek, the low wind's wailing dirge;
The arrowy prows, that swift as wild wings urge
    Their trackless way across the swelling floods,
Caves that yawn o'er leagues of wonders huge,
    And over all the endless solitudes
Of forests wide and dim—his own land's dark old woods!

SNOW SONG.

Come, beautiful snow, and bury the world,
    It is foul with many a stain;
And I long for a little while to look,
    And behold it clean again!

Restlessly heaving to and fro,
    Earth's millions turn and toil,
And lives are worn and traded away
    For a foothold on God's soil.

For the body's bread and its clothes,
    Its dwelling and burial place,
The soul is forgotten and thrown away,
    And leaves but dust in its place.

Evil is cruel, and fierce, and strong,
    And stealing the name of law,
Sends love to the gibbet and truth to jail,
    And the human to market like straw.

My soul rebels with a fierce revolt,
    But they bid it to bear the ill,
For a woman's sense of right they say,
    Must wait on another's will.

They bind the hands and gag free lips,
    And mock the exceeding pain;
Oh! the earth is foul! thanks, beautiful snow,
    For making it clean again!

MARTHA

I LOVE thee, Martha, with a love that liveth,
    With my deep passion for all lovely things,
Yet with that sad misgiving that receiveth,
    From ceaseless changes its melancholy tinge.
Love is our life—Earth's beauty—Heaven's brightness—
    The soul's responding to another's call;
The mountain's strength, the cloud-wing's fairy lightness,
    The psalm of waves, of wind, and waterfall—
Wake in our hearts that restless sweet emotion,
    That gusheth upward toward the bright and true,
Even as rise the billowy tides of Ocean,
    When Dian beckons from the empyrean blue.
But when a true heart, in its ceaseless beating,
    Throbs out warm comfort to those lingering near,
Though in a crowd it give its passing greeting,
    The warmth it giveth makes that true one dear!

STAR-BEAMS IN SHADOW LAND.

Day, crowned with sunbeams, sinks to rest,
    Pillowed 'mid clouds that blush and glow,
Transfigured by their royal guest
    From dun to rosy gold and snow.
The hand of evening comes to write,
    In shadowy letters on our wall,
Weird promise that the mist-veiled night
    On day's broad realm awaits to fall.

Yet ere that spell-enchanted light—
    That strange green light of summer eves,
That comes before the fire-flies light
    Their lamps among the whispering leaves—
Comes with its vague things half revealed,
    Like memories that we can't recall,
That tempt pursuit, but only yield
    An echo's answer to our call—
Let us put by each day-worn thought,
    And, in this Sabbath of the hours,
Wait for the good so often sought,
    So seldom found 'mong "human flowers."

That wisdom schoolmen have not learned,
    That deep-read sages oft deny—
By churchmen laughed to scorn, or spurned,
    Or passed in heedless silence by,
And yet which yearning spirits long,
    And hunger weary years to know,
Tho' sung in every wood-thrush song,
    And breathed thro' May woods' fragrant snow;
And this green forest twilight tells
    In sybil tones thro' dark-grouped trees,
One of the spirit's holiest spells,
    The key-note to its mysteries.

The heart in crowds which vainly sought
    Its secret pass-word answered back—
The one who learned what schoolmen taught
    To find, through his own soul, the track
By which those weird impressions come,
    Unasked, to haunt his thoughtful hours,
With visions of that Eden home
    The gypsies tell us once was ours—
Yet learned in vain; for through the soul
    No chart hath marked the devious way,
And, far as ever from the goal,
    He learned by turns to doubt and pray.
The one who sought for spirit lore,
    Of churchmen learned in creed and law,
Conning dry bones of dead faith o'er,
    Still doubting all its proofs he saw—
These might have learned, in this hushed wood,
    Alone with God and Nature's scroll,
In this still, solemn solitude,
    These untaught lessons of the soul.

Seek, delve and search, for wealth of gold,
    For knowledge of God's outward plan,
Toil for all learning, new and old,
    Make thought's rich spoil thine—if thou can;
But o'er the soul's entrenchments deep,
    No toil can bridge thy curious way;
To scale its ramparts, sheer and steep,
    In vain is art's or strength's essay;
For here, as by the Jordan's flood,
    Thy thought, like Israel's host, must stand:
Still the wild chafing of thy blood—
    Relax thy grasping, toil-stained hand—
Cool thy vexed spirit's feverish heat—
    Soothe down its each impulsive thrill—
Rest here, unshod, thy way-worn feet,
    Obedient to the words, "Stand still!"
Toil cannot win—gold cannot buy—
    Lore cannot give the soul its keys;
Itself sees not the seeing eye—
    Mind knows not mind's deep mysteries;

But like the glass reflecting back,
    The eye that seeks itself to see,
This twilight shows us what we lack,
    And what we are, and are to be.
And through the drifting mists that weave,
    Through warp of fact this woof of thought,
The waiting spirit can receive
    The truths its teachers have not taught,
And dimly as through distance far,
    Beholds its own deep secrets lie—
Or shadowy types of what they are,
    That vanish as they come more nigh.

And this is all—yet upward ever
    Those visions point us as they go,
Mute promise that past death's deep river
    Our souls shall learn themselves to know.

STAGNATION.

His spirit bears no weight of woe,
    Nor is it saddened by one pang;
His heart beats no despairing throe,
    No storm-clouds o'er his spirit hang,
Nor dimming tear-drops fill his eyes,
To tell of grief or sad surprise.

Yet pleasure hath but little part
    In the dull chaos of his brain,
Mirth quickens not his sluggish heart,
    Which only knows it feels no pain—
For joy hath lost the glad control
With which it ruled his willing soul.

His stagnant thoughts in silence creep,
    Nor seek new mysteries to explore,
Unheeding as the rocks that sleep
    On grim oblivion's murky shore;
Nor, but from habit, would they turn,
To where Heaven's watch-fires dimly burn.

The dormouse in its winter cell,
    Its death-like lethargy may break
In winter's wildest storm,
    As he from out this stupor wake,
While wintry circumstances hold
His spirit in their icy fold.

O, spring-time! hast thou lost thy way,
    And lingered in the sunnier bowers
Of hearts that catch each genial ray,
    To warm affection's springing flowers,
That thou hast left life's freezing chill,
Upon his wearied spirit still?

LULLABY.

O, WHY that sad, murmuring, half-conscious strain,
"Be hushed my dark spirit!" sung over again?
Are there storms with their shadows cast over thy sky,
That thus thou should'st sing them their lullaby?
Do the vapors of cloud-land thy spirit enfold
As the dun mists of autumn their volume have rolled
Round the mountain that basked in the summer sun
Ere the hours of darkness and gloom had come?
Or in that strain is there lingering still
Tones that no more thy soul shall thrill—
Echoes of voices that far through the past,
The bewildering spell of their melody cast—
That over, and over, and over again,
Dream-like and low, like the wind harp's strain,
"Be hushed my dark spirit!" seems ever to swell,
Unbidden from out of thy heart's deepest cell?

Is there dwelling a magic in sounds like these,
Which, faint like the voice of the dying breeze,
Comes over the heart in its loneliest hour,
And soothes it to rest with its quiet power?
The weary heart bowed with its burden stills,
As that mournful sound through its chambers thrills—
Stills from its throbbing tempestuous pain,
To list to that sad sound o'er again;
For weary and sad tho' the spirit be
It echoes the voices of memory.

SHEAVES OF TIME'S HARVEST.

TIME passed his hand o'er the brow of youth,
And ploughed deep furrows where once 'twas smooth,
Then he wrote great lines of thought and care
In the place of the smiles that it used to wear;
Then sprinkled the ebon locks with grey,
And faded the light of the eyes away,
And the reaper smiled at the mourners' grief,
As he gathered home this ripened sheaf.

Time stood by a forest dim and old,
As its thousand years were well nigh told,
And its fallen kings lay mouldering there,
Where the grey moss swung in the chilly air,
From ocean to ocean's distant shore
That trackless forest shadowed o'er;
But a nation toils where those wild-wood leaves
Were garnered once with that reaper's sheaves.

Time stood on Baalbec's giant walls,
And paced thro' proud Palmyra's halls,
And like the echo of his tread
Came funereal wailings for the dead;
And now the desert blasts alone
Sigh o'er each fallen monarch's throne—
The only spirit abroad that grieves
Over those long since gathered sheaves.

Time leaned against the humble shed,
That sheltered the starving peasant's head,
And though heart-strings broke as the roof fell in,
The groans of the dying were drowned in the din.
What recked old Time the peasant's woes?
He had looked on a dying nation's throes!

Alike to him man's joy or grief,
As he gathers one more human sheaf.

Yet waiteth not that reaper dread,
For the flower to wither and droop its head,
For he cuts with his sickle, sharp and keen,
The golden ear and the leaflet green,
The babe that sports at its grandsire's knee,
And the grey old man alike takes he,
The starting bud and the withered leaf,
He gathers to add to his well-grown sheaf.

The rose-tinged petals together rolled,
And the bud untwisting each fragrant fold,
The flower on its stem, scarce fully blown,
He bears with his sheaves to his harvest home;
The youthful while hope still paints on the air,
Visions of glory enchantingly fair,
The reaper gathers, nor heeds our grief,
He has need, perhaps, of this blooming sheaf.

Alone that silent reaper stands
As he binds his sheaves with his bony hands,
And he scans the field of his harvest o'er
As he scanned it a thousand years before—
And he laughs as he watches the puny toil
Of those whose labors he makes his spoil,
For the world and its creatures, like Autumn leaves,
He binds together—Time's harvest sheaves.

A VOICE FROM AFAR.

Christmas Eve, 1855.

I CALL thee! Dost thou hear my call?
　Or have those thrilling inner chords
Become too dull, and still, and cold,
　To vibrate to my words?
Within that chamber of thy heart
　Where old things half forgotten lie,
By dust and cobwebs wreathed and veiled,
　Comes no fresh life to memory?
Stirs not within that silent fane,
　Some pulse of life lived once before,
When well-known hands unbar its gates,
　And well-known footsteps cross its floor?
Send back those walls no echo note,
　To sounds that break their holy hush,
When once they pealed an answer back,
　To every wayward music gush?
Within that temple's inmost shrine,
　Will memory's statue breathe no more,
Like Memnon's, those deep anthem tones,
　It chanted oft in days of yore?

There is a charm to break its sleep,
　(Would I could wake that magic spell)
When gathering its long silent voice,
　Loud, deep, and long, its tones shall swell;

But till its long appointed hour
　Slow wheels around it must sleep on,
And though I rouse an echo's voice,
　I cannot wake one *living* tone.

I enter that deserted room,
　　My garments brush the dust away,
One quenchless votive lamp burns on,
　　To light my dim and dusky way;
Feels not that pallid statue's brow,
　　The living hand upon it pressed?
Does it not feel the holly wreath,
　　I bring to crown its silent rest?

I have been with thee—but I turn,
　　And bolt and bar that temple's door;
And know not if one life thrill told
　　Thy soul my shadow crossed its floor!

ANNIVERSARY LETTER.

CONDERSPORT, *Christmas Eve,* 1856.

I ONCE had a friend, who hath passed away,
Like the morning mists of a summer's day:
Forth like an anchorless barque he went,
But I know not whither his footsteps bent.
Is he wearily climbing the pathway of fame,
To hear echo mockingly shouting his name?
Is there glowing around him the sunshine of home,
Or treads he the "valley of shadows" alone?
'Neath a rainbow arch in the unknown land,
Enrol they his name in a white-winged band?

But a shadow falls on me with a menace of woe,
Ah! his soul is yet fettered with earth-bonds I know,
For never yet did that shadow fall,
From a spirit released from its earthly thrall,
Nor, alas! from a soul with its sins forgiven,
Spreading its wings for its home in Heaven!

Ah! Soul of my Friend! As the clouds in the sky,
Trail their shadows o'er all things beneath them that lie,
So the shade that cast over my spirit its chill,
Hath shown me thy soul is enclouded still!

But hast thou no token but this one to-night?
No message to speed like an "arrow of light,"
Athrill with thy feeling—aglow with thy thought?
Oh! woe for the lessons that world-craft hath taught!
Oh, woe! if the love which hath taught thee the cause
Of every effect of God's wonderful laws,

138

Hath curtained thy innermost self from the light—
If reason hath blinded thy instinct's sight,
And deeper woe, if the dross of earth—
Things meaner, and less, and of poorer worth—
Have darkened thy soul with their daily care,
Till thou seest but their mildew everywhere!

ANNIVERSARY LETTER.

LEXINGTON, ILL., *Christmas Eve*, 1857.

I HEAR the rush of mountain streams,
Underneath the evergreens,
   That bend beneath their load of snow;
I hear the north wind's echo call,
From mountain unto "mountain wall,"
   Where the dark pine woods grow.

I see the unfading mosses cling,
And hosts of yellow lichens spring,
   From leafless trunk, and fallen bough—
I see the green club-mosses creep,
Along the ferny ledges steep,
   As tho' I stood among them now!

The thousand miles that intervene,
Grow narrow as the pebbly stream,
   Before the homestead door;
And as night's deep'ning shadows darken,
I almost hold my breath to harken,
   To the footsteps on the floor;
And the old familiar call,
From the stairway in the hall,
   Sounds in my ear once more.

But I turn me slow away,
From the flickering lights that play,
   On the cottage window bars,
And a mist obscures my sight—
I may not pass home's door to-night—
   So reads the fiat of the stars.

Ah me! 'Tis Christmas Eve again,
And it belongs not unto them,
    But is like a monumental stone,
    Sacred to memory of one,
Passed and gone—I know not whither—
For a time—perhaps for ever!

'Tis Christmas Eve! I sigh with pain,
    For old Time's kaleidoscope
    Has not answered last year's hope—
Has not returned the lost again—
    Nor given me the faintest trace,
    Of the lost one's hiding place;
And these lines I now indite,
    By thy eyes may ne'er be read,
For I know not if I write,
    To the living or the dead.

But whether in the quick warm flesh,
    Or in robes that angels wear—
Whether 'mid old scenes or 'mid fresh—
    Or full of joy, or full of care—
If on a scaffold or a throne,
    Or if released from earthly thrall,
Thou hast learned 'mid stars to roam,
And claim yon over-arching dome
For ever as thy glorious home—
    I ask not; while the snow flakes fall,
Upon the hills we used to tread,
    Upon this evening I will call
On thee, if living—or if dead!

There! good night! God's blessing on thee!
    Blessed dreams attend thy sleep;
If in Heaven, may angels crown thee,
    And waft thee to an upper deep!

MODERN FAIRIES.

THEY say that the elves are dead and gone,
   And that never more by hill or lea,
Will the fairy rings in the grass be worn,
   By the nightly dance as they used to be.

The frost may blight, and the drouth may burn,
   And the grasshopper harvest the hay,
No more will you charge it to witch or elf,
   Since their games have had their day.

Believe it not! They are living still!
   You can find their traces night by night,
Where they spread a thousand gossamer webs,
   About in the dew to bleach them white.

They are spread like veils all over the grass,
   And over the drowsy red clover heads,
White nettings of mist, with a wealth of pearls,
   That are strung like beads on the shining threads.

Go, rob from the fairy bleaching ground,
   A veil to shadow some beaming face—
The pearls dissolve, and a knotted string,
   Of that fairy web is the only trace!

Ay! watch for the elves! They are living still!
   But go to the haunts where they love to go,
Where the red moss chalices left behind,
   Are telling of last night's nectar flow!

Await them in faith! await them in love!
   When the young moon's boat sails down the west,
And the blessings they leave in the woods and fields.
   Will linger around you and make you blest!

FRAGMENT.

The winds may give the treasures back,
    They gather from the dewy flowers,
Leaving along their fragrant track,
    Sweet perfumes from Arcadian bowers.

The waters may return the bread,
    The liberal hand upon it cast,
But when will Time's returning tread,
    Bring back the dead, the buried past?

Morn may restore those glowing hues,
    That evening's curtain wrapped in shade,
And evening shed again those dews,
    That morning dried from wood and glade.

But Memory with oft wearied feet,
    Must wander o'er past scenes in vain,
For never from their dark retreat,
    Will bygone hours return again!

VIGIL LESSONS.

A STUDENT marks each moment as it passes,
    By the slow droppings from the eaves,
Pattering on the broken, rustling grasses,
    And the dock's red-spotted leaves.

Then out so solemnly and slowly,
    The drops the precious moments dole,
That the dim night hours, calm and holy,
    Seem endless to that watching soul.

While day gathereth fresh splendor,
    In the caves of eastern seas,
Earnest night, so calm and tender,
    Locks sleep's gates with silence' keys.

But past their portals ere their closing,
    A horde of spirits had entered in,
That 'mid the body's sweet reposing,
    Their pranks upon the soul begin.

Hope, who cheered thro' day's hard battle,
    Queens it now on reason's throne,
The real's work day toil and rattle,
    Merged in a music all her own.

Or dreaming Faith o'erleaps for ever,
    Forgetful of life's parting pain,
That deeply shadowed, silent river,
    Once crossed, which ne'er is crossed again.

Or stricken grief, with icy fingers,
    On the heart's bare quivering chords,
Plays wailing music such as lingers,
    In love's despairing, farewell words.

But oftener fancies fast and fleeting,
    Wild revels hold in every dream,
Like troops of summer lightnings meeting,
    Kindred flashes in each stream.

So pass the night hours with the sleeping,
    But without the gates of rest,
Unwilling, weary vigils keeping,
    Sits wakeful thought's unsleeping guest.

Then that student still and lonely,
    Learns of voiceless Night and Thought,
For in the deepest silence only,
    Their higher miracles are wrought

On the beating pulse of nature,
    Lays each witch her finger light,
Revealing by that secret feature,
    Every heart-throb hid from sight.

It is worth this watch unsleeping,
    With the rain-song in one's ear,
Thus to gather to one's keeping,
    Earnest truths taught only here!

LIFE'S NOONING SONG.

THANK God! I'll be a child no more!
    Life's earlier woes and fears are dead,
I've trod upon youth's utmost shore,
And life full grown is half way o'er,
    With all its storms and darkness fled.

My morning dawn was dark and wild,
    But cheery grows the lengthening day,
And since I've ceased to be a child,
The genial skies have ever smiled,
    And now I love its lingering way.

Two brimful cups life offered me,
    A cup of joy—a cup of care—
My right hand grasped the draught of glee—
The left thrust care's grim bowl from me—
    And life has since grown rich and fair!

It may be dangers all unseen,
    Have lain in wait about my path—
It may be pitfalls lay between
My footsteps in the path I've been;
    What matter? I have passed unscathed.

It may be down the hidden slope,
    My restless feet must soon descend,
I must with fiercer dangers cope—
I will not fear, but go in hope,
    And trust in the Unfailing Friend!

THE MOON OF BLOSSOMS IN PRAIRIE LAND.

ALL the woods are full of blossoms,
It is now the Moon of Blossoms,
The Moon of all the snow white flowers.*
The May cherry has departed,
Swung its feathery plumes of sweetness
For awhile on every wind breath,
On the slender twigs that bore them;
Then upon the breeze that kissed it
Strewed the snow flakes of its petals.
    The wild plum had shed its sweetness,
Till the air was laden with it;
Till the drowsy air grew drunken
With the fragrance of its blooming.
Then the blossoms blushed with pleasure—
Blushed to find themselves more fragrant
Than all other early flowers.
    And the wild bees from the torpor,
Of the long and idle winter,
Waked to work amid the flowers.
Then from out the leafless snowdrift,
From the plum-tree's blushing snowdrift,
Came the busy hum of labor,
Came the happy song of labor,
Mingled with the drowsy odor
Of its own exceeding sweetness:
Till its petals withered slowly,

*The Senecas call May the Moon of White Blossoms.

147

Shrunk and shrivelled in the sunshine.
They refused to leave the calyx,
And go drifting in their beauty,
Like the apple's faded petals;
Like the graceful snow flakes drifting
In the passing winds that shook them.
Then the woods ablaze with purple,
Flushed and flamed in all their borders.
Crimson fringes edged the prairies,
Rosy purple framed the rivers
With a setting richer, rarer,
Than the costliest palace mirrors,
Though the lowliest and poorest
With their seedy rags unchidden,
And their bare feet unforbidden,
Might adjust their shaggy tresses
By those grand and wondrous mirrors;
And the children gathered treasures,
Priceless treasures, from the tree-tops,
In those countless purple branches,
In the pea-bloom of the red bud.
    But before this flushing fadeth,
Comes the dog-wood's robes of whiteness,
Mingled with this royal raiment,
And from out each clump and thicket,
And from every patch of brushwood,
From beside the pools of water,
From the hillock and the sand-ridge,
And from out the isles of timber,
And the borders of the prairie,
Comes a fragrance fresher, sweeter,
Than the breath of strawberries,
Than the odor of the roses;
Like the smell of fruit and flowers,
With a fairy's blessing on it,
Making it a thing unreal,
Only something we have dreamed of,
Though it freshens all the senses
With its penetrating presence.
'Tis the fragrance of the crabtree,
'Tis the bursting of the flowers,

From the rosy buds, that swelling
With the prisoned odors in them,
Could not longer hold their riches.
　　And the earth is strewed with blossoms,
Blue as is the sky above it.
Blue-bells blossom by the water,
May-bells bloom among the grasses,
The wild larkspur seeks the road-side,
And the blue phlox everywhere
Utters forth its soul in odor.
　　And like stars amid the azure
Early buttercups are blooming;
And this bursting world of flowers,
Keeping time to all the trilling,
Sinking, swelling songs and carols,
Of the whip-poor-wills and robins,
Swing and sway upon the breezes,
In the airy dance of flowers,
In the white-robed Moon of Blossoms.

A REVERIE.

A CLOUD o'ershadoweth my soul,
 With many a silvery purple fold,
 And ragged edge of hazy gold,
And sunset smiling through the whole.

A shadow, in whose softening shade,
 The long-ago comes floating back,
 With song and perfume on its track,
And seems as 'twere the present made.

The far-off, through the wilds of space,
 A welcome vision draweth near,
 Wreathed with familiar smiles, and dear,
To me in this lone stranger place.

The future, in a misty veil,
 Draws this magic cloud beneath,
 Like Dian in a vapory wreath,
And tells her prophet tale.

I seem to leave this mould of clay,
 And out among these things of thought,
 So near me by enchantment brought,
To go, ethereal as they.

I thank thee, cloud, that thou hast made
 Amid life's toils, so stern and real,
 An hour so tender and ideal
As this, of light embalmed in shade.

I thank thee, that the glare of day
 Sometimes dissolves in clouds like this;
 Where rainbow prophecies of bliss,
Shine on my often darkened way.

ALTER EGO.

We meet as the waters meet,
   Our souls together run,
   And mingle into one,
And life grows deep and sweet.
   From the heart that beats in thee
   Stirs the inmost pulse in me,
And the thought that in my brain,
   Struggles faintly to be said,
Thy soul echoes back again,
   Ere a sentence has been made.
We quicken each the other's thought,
   But so akin our sayings rise,
   That be they vain, or be they wise,
Which spoke them we remember not.

The streams of our lives diverge, and change,
   And labor attends us whither we go;
New events and faces, diverse and strange,
   Drift around us, wavering to and fro;
And each hath its load of especial care,
And a different tide, a different air,
   Is urging us evermore on and on.
But we meet again, as the waters meet,
   And the past is vanished away and gone;
And now life has grown—oh, more than sweet!
The treasures we've gathered, the lives we've led,
Have deepened and widened the heart and head;
And yet, ever tendeth my soul to thee—
And ever thy spirit is coming to me,
As the rivers tend down to their home in the sea!

THE END

Nov. 3 1860

AFTERWORD.

Laurie Lounsberry Meehan

Alfred, 1850–1855

In an 1890 reflective essay on Elizabeth C. Wright, Mary E. H. Everett described her as a "teacher, friend, inspirer" and quoted her as saying: "To see the *real me*, you must go to Alfred when I am there. . . . There they call me 'Libbie,' and my life expands and puts on softness and completeness, just like my name!"[1] Much of Wright's success can be attributed to the support, teaching, and overall environment in which she was immersed at Alfred University in the mid-nineteenth century. As proscribed in its charter, the Alfred Select School opened in 1836 with "equal rights and privileges" for both men and women, with the intention to provide an education beyond the local district schools. It soon grew to become the Alfred Academy in 1843 and was incorporated as Alfred University in 1857. Due to the liberal nature of the townspeople and to strong leaders like William Kenyon, Jonathan Allen, and Abigail Allen, the students were exposed to unusually early support for women's rights and equality for all people.

Beginning with her first classes in 1847 through the culmination of her graduation in 1855, Wright found herself surrounded with like-minded souls who, as professor and later friend Abigail Maxson Allen declared at the 1873 Women's Congress in New York City, were willing to "be radical, radical to the core."[2] The relationship between the two women likely started in 1845 when Allen taught public school in Wright's hometown of Ceres, New York. In her few remaining letters, Allen inquires about "Libbie" in correspondence with

---

1 Mary E. H. Everett, "A Friend and her Letters," *Alfred University* 3, no. 2, November 1890, 10.
2 Allen had been invited by Julia Ward Howe to deliver the speech, which argued for coeducation. Abigail Allen, "Co-education," *Alfred Student* 1, January 1874, 1–3, and *Alfred Student* 1, February 1874, 13–15.

Everett,[3] cousin to Wright's husband Lyman Jewell (who attended Alfred in 1864). For their first few years, Wright and Allen were student and teacher but the slight difference in their ages, less than three years, allowed them to become lifelong friends and equals. While not of the same denomination, their respective upbringings in Quaker and Seventh-Day Baptist families laid the foundation for similar values and beliefs. The July 1858 camping trip described in Wright's essay "Into the Woods" had such a lasting impact that Allen included it in the memoir she wrote about her husband, Jonathan, after his passing in 1892.

Located in the small town of Alfred, situated in Allegany County in southwestern New York, the university's foundations and faculty mirrored the liberal attitudes of the townsfolk, most of whom were members of the Seventh-Day Baptist Church, a small but active sect on reform issues of the day. Like Wright, many of the residents were influenced by the social reforms of the "Burned Over District." It was a vibrant, progressive period in Alfred during Wright's time:

> Although the Southern Tier of New York was settled late, was still considered frontier in the 1830s, and ... remained isolated from contemporary excitements, Alfred's students were vigorously interested in reform and revivalism. While Alfred maintained staunch adherence to Seventh Day Baptism (resisting millennarianism, Mormonism, and utopian experiments), religious enthusiasm was high, revivals flourished, and social reform movements were well known here. Temperance, anti-slavery, and soon the issue of women's rights were of continuing interest to faculty and students.[4]

As most of the early settlers were Seventh-Day Baptists, they brought their ideals and common religious practices to Alfred: ideals that strongly influenced the development of the town for over one hundred and fifty years. Their Saturday Sabbath-keeping meant that businesses were closed on Saturdays and open on Sundays. They were staunch supporters of abolition and worked to bring an end to slavery by vocally opposing it (a resolution was passed against it at their 1836 General Conference), participating in the Underground Railroad, and volunteering to fight in the Civil War. After the Fugitive Slave Act was passed in 1850, sentiment in Alfred heightened to action as a number of residents (including students and alumni) moved

3 Abigail Allen Letter Collection, University Archives, Herrick Memorial Library, Alfred University.
4 Susan Rumsey Strong, *"The Most Natural Way in the World": Coeducation at Nineteenth Century Alfred University* (PhD diss., University of Rochester, 1995), 92.

to Kansas in an effort to influence the state's outcome as a free state. They were also much more open in their attitude toward women's rights than many others. While many of the residents and faculty were of the Seventh-Day Baptist denomination, the university was nonsectarian. The 1855 college catalog describes the "moral and religious influence" of the school by saying, "The members of the Board of Instruction belong to different religious denominations, and adopt an enlightened religious policy."

Past scholarly research by Kathryn M. Kerns and Susan Rumsey Strong on the university's early history support the influence of rural values and agricultural families as a reason for stronger support of women's rights:

> Among the forces that coalesced at Alfred to help shape the form of education for women before the Civil War were the personalities and values of teachers like the Allens. Other influences included the Seventh Day Baptist religion, the local community, the socioeconomic backgrounds of the students; and, these, in turn were affected by larger developments such as general social and economic conditions in the nation, the growth of new career opportunities for women and an increasing national commitment to education. To single out any one factor as dominant is to distort the picture and miss the complicated interconnectedness that provided much of the school's strength in the antebellum period.[5]

In the mid-nineteenth century, 40 percent of the student body was female, most coming from Allegany and adjoining Steuben County where the "majority of Alfred's students came from farms where women's work still played an important part in the farm economy. Daughters coming from such farms would be much more likely to view the family as a production unit and work as compatible with marriage, even after their schooling had taught them to strive for careers beyond the farm."[6] Wright's graduating class consisted of twenty-three students, a dozen of which were female.

In addition to the mix of genders, it was also not uncommon to find international students, further strengthening the sense of acceptance and exposure to world culture. During Wright's time, students in attendance came from varying locations such as Spain, Cuba, and France. Graduating with Wright was a young woman from Haiti, the first black student to attend Alfred. As a result

5   Kathryn M. Kerns, "Women at Alfred: Pioneers of American Coeducation," in
     Sesquicentennial History of Alfred University: Essays in Change, ed. Gary S. Horowitz
     (Alfred, NY: Alfred University Press, 1985), 58.
6   Kathryn M. Kerns, "Farmers' Daughters: The Education of Women at Alfred Academy
     and University before the Civil War," *History of Higher Education Annual*, 1986, 25.

of the famed 1858 camping excursion immortalized by Wright in *Lichen Tufts*, students (men and women) from the Seneca Nation reservation in nearby Salamanca, New York, attended the university as well.

The opening of the Erie Railroad in 1851 had a major effect on day-to-day life in Alfred. Agricultural goods were now easily shipped to larger markets and travel between cities was opened up as passenger trains ran the lines. Small and large businesses flourished in the town during the nineteenth century as did the educational system. The 1850 census shows there were 2,679 people living in Alfred, including almost six hundred born in Ireland (primarily railroad workers and their families, a history of strife in its own way). Facilitated by easier transportation, travel to rural Alfred opened the opportunity for national figures to visit, including Frederick Douglass (one visit was while Wright was a student), Julia Ward Howe, Sojourner Truth, Susan B. Anthony, and Elizabeth Cady Stanton, to name a few. It also allowed travel from Alfred to cities like Rochester, New York, where university president William Kenyon and the Allens attended the 1853 Teachers Convention, watching transfixed as Susan B. Anthony created a major stir by daring to speak up (the first time for a woman). Abigail Allen had ongoing interactions with Anthony and Howe throughout her lifetime.

Trained at Union College, Kenyon was instrumental in creating the foundation for the school and its ideals. Arriving in 1839 as the third teacher at the Select School, he was critical of "superficial women's education and worked hard to found a coeducational college at a time when women's higher education was still controversial and coeducation rare."[7] He dedicated his life's work to the school's success. Before his death in 1867, Kenyon served as teacher, principal, and university president.

Unlike many other schools of the time, the focus of the early curriculum was on teacher education rather than a higher education for ministers, partly in response to the high demand for teachers during the common school movement of the time. This was especially beneficial for women since teaching was viewed as a respectable and acceptable profession for them. Enrollment grew from the first class of thirty-seven students in 1836 (twenty-two were women) to just over three hundred students by the time Wright attended.

Discussion and action on the social reforms of abolition, temperance, and women's rights were supported and encouraged in Alfred. Most students were active in the student lyceums, which ultimately were training grounds for the development of critical thinking and public speaking skills: "Immediately after being hired at Alfred Academy in 1846, Abigail Maxson [Allen] took

7   Susan Rumsey Strong, *Thought Knows No Sex: Women's Rights at Alfred University* (Albany: State University of New York Press, 2008), 74.

an important and unusual step—she started Alfred's first literary society for women, the Alphadelphian Society. This was one of the earliest women's literary societies in the country, a crucial forum for debate, development of self-confidence, and encouragement to take an active, public role in society."[8] The role of the lyceums cannot be overstated. They provided a place for the development of public oration and debate skills not provided elsewhere for women at the time. Wright was a member of the Alfredian Lyceum, a later iteration of the original Alphadelphian Society. Support came from the local community as well, many of whom attended the annual public sessions of the lyceums held in conjunction with graduation. In 1850, at a dinner after commencement, various toasts were offered by faculty and community members, including one to the "Ladies of our Literary Institutions—may they ever spurn the rule of fashion, and be true and zealous reformers."[9]

Exemplifying that nature, student Myra McAlmont wrote home to her mother in 1852 saying, in part,

> I am studying Latin, Rhetoric, Algebra, and Analytical Geometry. The reason of my taking the last study which does not come in this course, was that about ten of our best scholars in the Institution, all gentlemen[,] formed a "crack" class. Miss L. Pickett and myself went in to show them what we can do. Thus far we have sustained ourselves with honor. You will perhaps be surprised that I should commence Latin. But I think it will do much to strengthen my mind if nothing more.[10]

McAlmont carried that high ideal and interest as she went on to be a professor of mathematics at the Female College in Little Rock, Arkansas. In her essay "The Nature Cure—for the Body," Wright decries the current state of women's fashions as well as the need for preventative medicine. In her letter, McAlmont mentions wearing "the Bloomer costume" the same time homeopathic doctors in Alfred (Hiram Burdick and Mary Bryant) were pushing for the (literal) loosening of women's clothing along with a more holistic approach to living through diet and exercise. Burdick, also a temperance lecturer and Seventh-Day Baptist minister, later married his fellow physician Bryant.

The first female lecturer, activist Elizabeth Oakes Smith, was invited to campus in 1854 by the Ladies Literary Society. The same year, female graduates began giving orations during commencement rather than just reading their

8 Strong, *Thought Knows No Sex*, 63
9 "The Fourth at Alfred," *Sabbath Recorder*, July 25, 1850, 21.
10 Myra C. McAlmont to Mrs. S. McAlmont, August 27, 1852. Letter in University Archives, Herrick Memorial Library, Alfred University.

work, the earliest known instance in the country. Oberlin, the first coeducational school in the United States, wouldn't allow women to even read aloud their papers; rather, a male professor did it for them. Witnessed by Jonathan Allen, a fellow student at the time, Lucy Stone protested this in 1847 by refusing to graduate. He was asked by Oberlin president James Fairchild how the issue was handled at Alfred and captured the Alfred sentiment concisely by replying, "The most natural way in the world. If a young woman is capable of writing a paper, she ought to be able to read it."[11] It was also in 1854 that Allen started teaching the "Legal Rights of Women" class at Alfred. A native of Alfred and student in the first class of the Select School, Allen joined the faculty in 1849 after completing his degree at Oberlin. After Kenyon's death, he was elected to lead the university as its second president.

Jonathan and Abigail Allen, first as students, then as professors and leaders of the university, exemplified the strong social reformers of the nineteenth century and emboldened their students to follow suit as they left Alfred to begin their own lives. Immersed in this environment, so closely matched to her own upbringing, Wright grew and thrived and followed the examples of her mentors to be independent, bold, and to speak her own mind.

Working in partnership, the Allens often lectured and wrote on their views of equality, including coeducation and women's suffrage. In an essay focused on rights, Jonathan Allen stated, "Being human, equalities or inequalities of faculties have nothing to do with rights. In the presence of rights, race, sex, or color distinctions disappear. The humblest and feeblest being has the same rights as the most powerful and gifted."[12] Another time he wrote, "All isolation of individuals, all segregation of classes, on the principles of caste, birth, sex, or occupation, becomes abnormal, dwarfing and distorting. . . . The essential powers of the spirit are neither masculine nor feminine, but human, sexless. Thought knows no sex."[13] Similar sentiments are expressed by Wright in her "Into the Woods" essay: "I wondered, then, more than ever, where people ever get the absurd notion of talking about 'refined' and 'vulgar,' or 'masculine' and 'feminine' employments. It sounds ridiculous as the French way of calling knives masculine and forks feminine. My knives are no more masculine than my forks. Elvira's shooting was as feminine as her curls, and the Professor's [Jonathan Allen] cooking as manly as his beard."[14]

11 Abigail Allen, *Life and Sermons of Jonathan Allen, President of Alfred University* (Oakland, CA: Pacific Press Publishing, 1894), 51–52.
12 Jonathan Allen, "Suffrage," *Alfred Student* 4, no. 7, April 1877, 74.
13 Jonathan Allen, "Mixed Schools," *Alfred Student* 4, no.1, October 1876, 2.
14 Elizabeth C. Wright, *Lichen Tufts, From the Alleghanies* (New York: M. Doolady, 1860), 51.

Not only were women admitted as students, they were also hired as faculty. In its first three decades, eleven women are listed as "professors" out of the total twenty-nine faculty names listed in the college catalogs. Fifteen women are listed as "teachers" in the same period along with twenty-five men. As one of those faculty, Abigail Allen was

> a strong-minded woman, dedicated teacher, and reformer who became an early suffragist, she was crucial in shaping the school's development and intensifying its liberal environment.... [She] made the strongest connection between educational ideals and social reform. Ambitious, intelligent, and hardworking, she pressed for social justice, a public role for women, broader employment opportunities, and egalitarian gender relations. In concert with her husband, she defined a uniquely liberal educational environment.[15]

In addition to their teaching and support of reform movements, the Allens were highly interested in natural history and spent much of their free time trekking around the countryside to find new specimens (rocks, fossils, plants) to add to their personal collection. Their love of nature extended to the aesthetics of campus as well. Similar to Wright, transcendentalist thought is reflected in their work as well; it was not unusual for Jonathan Allen to lecture on the topic of beauty and its importance in life. He felt that too often students were forced to learn lessons

> in little, low, half-made, rickety old buildings . . . [experiencing] many a dull, tedious school day . . . is whiled away, wherein the hungry soul of childhood is far away, listening in fancy to the merry chatter of the brook. . . . An ideal school is a home where hungering and thirsting souls are satisfied, where dormant energies are aroused, stimulated, inspired to noble life and action, where spiritual growth, strength, harmony, and beauty are the results; in short, develop all that is desirable to appear in the future.[16]

Their love of exploration was certainly shared by the other companions on the camping foray in 1858. Equally like-minded in so many respects meant that not only Wright celebrated the chance to accompany the Allens but so did fellow students Elvira Kenyon, Susan Maxson, and Weston Flint.

---

15 Strong, *Thought Knows No Sex*, 61.
16 Allen, *Life and Sermons of Jonathan Allen*, 67.

Kenyon (no relation to William Kenyon), member of the class of 1859, boarded with Wright in 1850 and had just completed her third year in the college course before the trip. While attending classes she also taught Latin as an adjunct teacher starting in 1857 (adding German instruction in 1862). She and Wright joined forces once again during the 1862 Ladies Literary Society Anniversary Session where Kenyon lectured on Elizabeth Barrett Browning and Wright read a poem. When the first college degrees were awarded in 1859, both Kenyon and Abigail Allen were recipients. In addition to teaching, Kenyon was also known to give campus lectures and critical scripture analyses during the Seventh-Day Baptist annual conferences. She left Alfred in 1866 to become the head of the Seminary for Young Ladies in Plainfield, New Jersey, receiving an honorary degree from Alfred in 1875.

Maxson (no relation to Abigail Maxson Allen) had just completed her fourth year of the teaching course when she signed up for the hiking trip. The following year, she married Rev. Stephen Burdick, a Seventh-Day Baptist minister who had graduated from Alfred in 1856.

Flint hailed from Great Valley, New York and had recommended the location for the 1858 trip, finishing his teacher's degree at Alfred just a few weeks before they embarked. He received his master's degree from Union College in 1863 and "true to the teachings of Alfred's anti-slavery spirit, as he had helped to send arms to John Brown in Kansas, so in St. Louis he was a somewhat active member of the underground railroad, helping slaves over into Illinois."[17] Among other accomplishments, he served as editor and publisher of the St. Louis *Daily Tribune*, as a US consul in China, on the Republican National Committee, and received a law degree from Columbia University. Later, he was librarian of the Scientific Library for the US Patent Office, statistician of the US Bureau of Education, and librarian of the public library in Washington, DC, where he lived for many years. Alfred University bestowed two honorary degrees on him: doctor of philosophy in 1886 and doctor of laws in 1900.

Numerous other students from the same time period left Alfred and made remarkable impacts. Sarah Langworthy graduated from the Alfred Academy in 1851 and later ran a sanitarium in New York City with her husband Dr. George Taylor. Part of the Spiritualism movement of the time, they cared for and recorded the séances of noted medium Katie Fox[18] and treated many high-profile individuals, including the Beecher sisters, Abby Hutchinson, Elizabeth Cady Stanton, Olivia Clemens (Mark Twain's wife), and P. T. Barnum.

17 "Weston Flint, LL.M., Ph.D.," *Alfred University* 4, no. 3, February 1892, 17.
18 The Taylors interactions with Fox are recorded in W. G. Langworthy Taylor, *Katie Fox, Epochmaking Medium and the Making of the Fox-Taylor Record* (New York: G. P. Putnams' Sons, 1933).

Mary Willis attended the Alfred Academy at the same time as the four students on the 1858 trip. She housed and helped escaped slaves reach Canada as they came through Town Line, New York, a town that seceded from the Union during the Civil War.[19]

Hannah Simpson, who received her medical degree and was an ardent abolitionist, mentions in her 1861 diary that she had breakfast with Wright. Multiple other women earned their MD degrees after attending Alfred in the mid-nineteenth century, while numerous others went on to teach, do missionary work, or, like Wright, work toward supporting various reform movements. Many did so while also married and raising families. Such was the foundational support and education they received from the institution and community found in Alfred, New York.

19 Willis's story is highlighted in Daren Wang, *The Hidden Light of Northern Fires* (New York: St. Martin's, 2017), whose family owned her house while he was growing up. Willis attended Alfred University in 1858–59; her two brothers also attended a few years earlier, crossing paths with Flint, Kenyon, Maxson, and Wright. While the book is historical fiction, it represents the general story line of Willis.